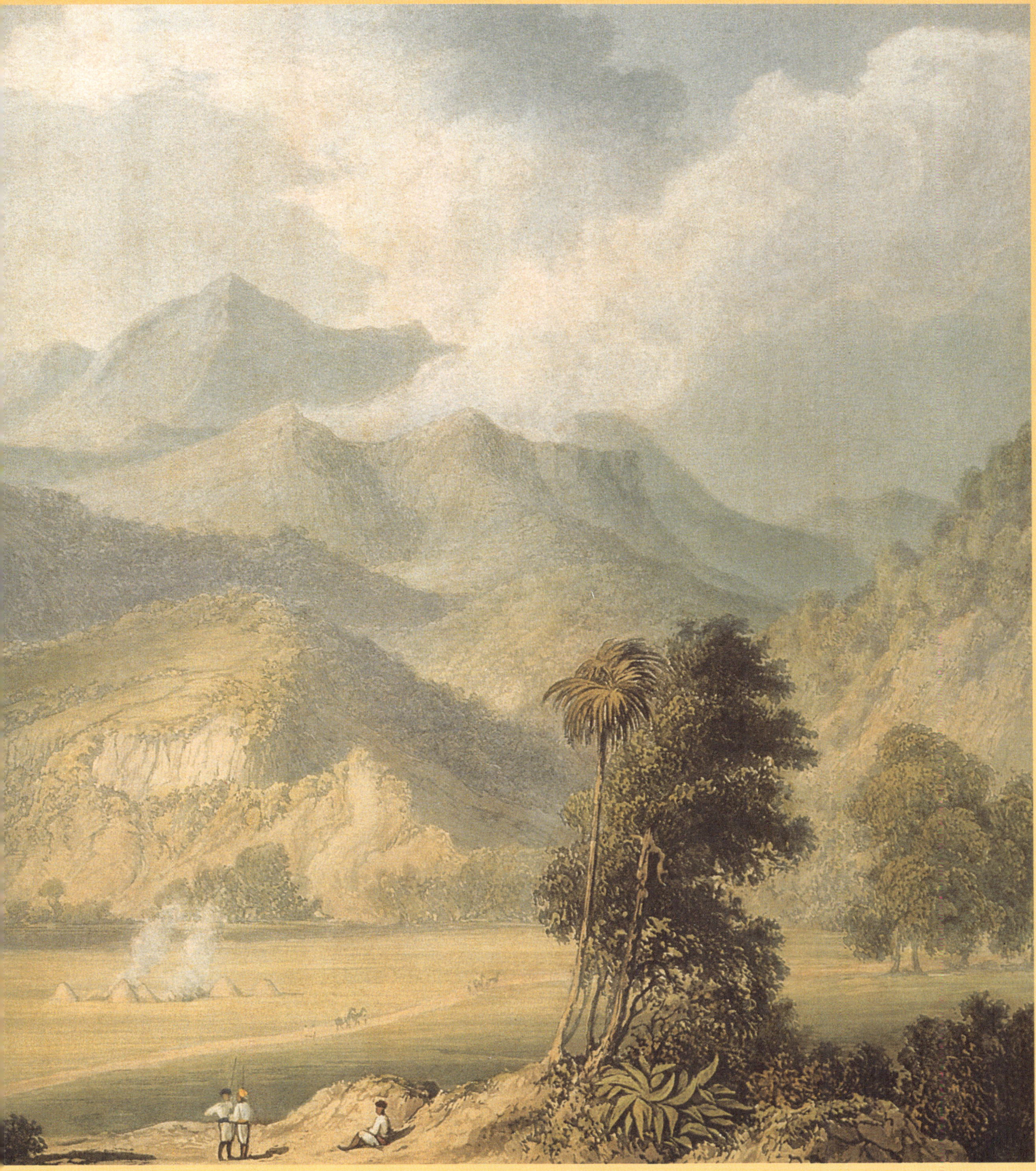

# TREASURES OF INDIAN WILDLIFE

# TREASURES

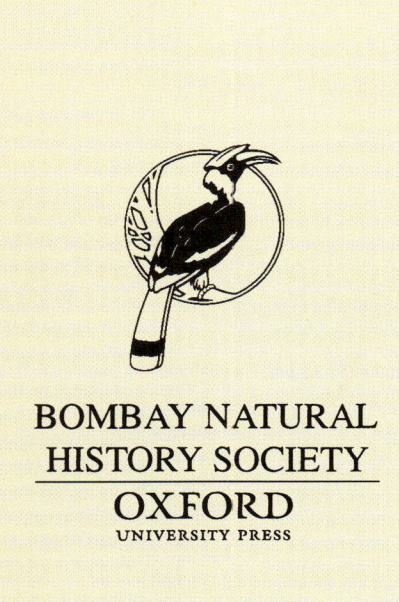

**BOMBAY NATURAL HISTORY SOCIETY**

**OXFORD** UNIVERSITY PRESS

# OF INDIAN WILDLIFE

*edited by* Ashok S. Kothari and Boman F. Chhapgar

*Advisory Committee* Asad R. Rahmani · Rachel Reuben · Rishad Naoroji · Gayatri Ugra · Isaac Kehimkar
*Technical Guidance* M.R. Almeida · J.C. Daniel · Naresh Chaturvedi · Varad B. Giri · William Connal
· S. Balachandran · Ranjit Manakadan · Vibhu Prakash
*Fund Raising* Pheroza Godrej · Dilnavaz Variava · Asheesh Pittie · Kavin Parikh · Radhika Sabavala
· Ruby Madan · Deepak Apte · Hansa Kothari · Rudra Connal · Nikin Mehta · Raj Shekhar Parikh
· Premal Parikh · Rajal Kapadia · Sheetal Shah · Mamta and Zubin Mehta · Ratnesh Desai · Prashant Mahajan
*Coordinators* M.G. Mathews · Joan D'Souza · Sanjay Sarange · Sachin Kulkarni · Shabnam Shaikh · Deepali Chaubal
· Santosh Mhapsekar · Sunil K. Ghanvilkar · M. Narayan · Dipti Sawant · Trupti Chavan · Shubhangi Vedak
· Mangesh Ghodke · Uma Pratap Singh · Leela Daniel · Vasant Rama Naik · Suresh M. Shetty
· N.V. Dighe · Prashant Kumbhar · Shruti Paradkar · Jaya Gumala
*Secretarial Assistance* Jayaprakash K. Menon · V. Gopi Naidu · Shalet Alva · K. Haridas
*Attendants* Tarendra Singh · Sadanand Sakharam Shirsat
*Photography* Dharmendra Mistry

© Bombay Natural History Society, 2005
ISBN: 019 567728 5
Price: Rs 1900.00

*Produced by* **Marg Publications** for the Bombay Natural History Society
*Editorial* Savita Chandiramani · Rivka Israel · Arnavaz K. Bhansali
*Design* Naju Hirani
*Production* Gautam V. Jadhav · Vidyadhar Sawant
*Data Entry* Rajkumari Swamy

*Processing at* Reproscan, Mumbai 400013.
*Printed at* Infomedia India Limited, Nerul, Navi Mumbai 400706, India.

Published by Rachel Reuben, Honorary Secretary, The Bombay Natural History Society, Hornbill House,
Dr Sálim Ali Chowk, Shaheed Bhagat Singh Road, Mumbai 400023
And copublished by Manzar Khan, Oxford University Press, YMCA Library Building,
Jai Singh Road, New Delhi 110001.

Oxford University Press, Walton Street, Oxford OX2 6DP
Oxford, New York, Auckland, Bangkok, Buenos Aires, Cape Town, Chennai, Dar es Salaam, Delhi, Hong Kong, Istanbul,
Karachi, Kolkata, Kuala Lumpur, Madrid, Melbourne, Mexico City, Mumbai, Nairobi, Sao Paulo, Shanghai, Taipei, Tokyo, Toronto

No part of this publication may be reproduced or stored in a retrieval system, or transmitted,
in any form or by any means, without the prior written permission of the copyright holders.

To the memory of the now extinct
INDIAN CHEETAH
and
with concern for the many species
of Indian wildlife
under threat of extinction

"**Cheetah**". Drawing by H. Weir. *Routledge's Picture Natural History* by the Rev. J.G. Wood, engraved by the Dalziel brothers, 1885.

# Contents

**Foreword** B.G. Deshmukh .................... 10

**Preface** .................... 12

**Acknowledgements** .................... 13

**The Founders of the Bombay Natural History Society** ........ 14

**Selections from Books, Journals, and Gazetteers**

Banian Tree, the Pride of Hindostan .................... 17

The Manpoora Tiger (About a Tiger Hunt in Rajpootanah) .................... 22

Baug! Baug! Baug! .................... 30

Gond Fable of Singbaba .................... 35

The Oppression of Man .................... 41

A Forest Fire .................... 46

Tigers of Salsette Island .................... 52

Goruckpore Terai, where Tigers were "Plentiful as Blackberries" .................... 55

Wounded Tigers, Panthers, and Bears .................... 60

The Indian Wild Boar as Grazier .................... 65

Khedda, Elephant Trapping .................... 68

Lion-Hunt in Hurriana .................... 72

The Past and Present Distribution of the Lion in India .................... 76

An Adventure with a Cobra de Capella .................... 82

The Poisonous Snakes of the Bombay Presidency .................... 86

A Giant Squid Attacking a Ship .................... 94

Wildlife around Cambay and Ahmedabad .................... 98

Sport and Natural History around Deesa in Northern Gujarat .................... 102

Vanishing Wildlife of Panchmahals .................... 112

Trees and Plants of Rewa Kantha .................... 114

Khandesh, a Stronghold of Wild Beasts ....................................................... 120

Wildlife of Jalpaiguri District ................................................................... 125

A Visit to Narcondam ............................................................................... 130

**Gleanings from Miscellaneous Notes in the *JBNHS*** ............ 137

The Journal: Its Role in Indian Natural History

* Red Ants' Nests * Panthers Tree'd by Wild Dogs * How the Monitor Defends Itself * Taming a Heron * The Sagacity of the Langoor * An Aggressive Cobra * Wolf Cubs * Muscular Action after Death * A Plucky Instance of Panther-Killing * Red Ants as Smelling Salts * The Young of the Hunting Leopard * A Centipede Eating a Snake * A Porcupine-Tiger Tragedy * The Small Indian Mungoose * Mongoose v. Cobra * Catching a Cobra with Bare Hands * Capturing Tigers with Bird-lime *The Tiger and the Train * A Bird Passenger on a P.&O. Liner * Python and Monitor * Python Attacking a Spaniel * Tiger Killed by a Cobra *A Panther Shoot at Sea * An Example of an Assisted Passage * The Hunting Leopard in the Central Provinces * How the Monitor Lizard Sits in its Burrow * "Feline Government Gazette" * Tigers Swimming * Eleven Koel Eggs in a Crow's Nest * Remarkable Behaviour of a Tigress * A Newly Born Bison Calf * Wild Dogs Killing a Panther * The Ashoka Tree *A Leopard-like Tigress * Smoking a Panther to Death * The Mating of Elephants * The Hatching of a Mugger * The "Courtship" of the Monitor Lizard * Rarity of Man-eating Tigers in South India *Fishing with the Indian Darter in Assam * "Pandadi or *Strobilanthes callosus* in Junagadh * The Flowering of *Strobilanthes* * The "Dew-Claws" of the Hunting Leopard * Rabies in Tiger * Implanting African Lions into Madhya Bharat * Footprints of "Snowman"

**Bibliography** .............................................................................. 210

**Sponsors** ................................................................................... 213

# Colour Plates

BLACK-CAPPED KINGFISHER
(*Halcyon pileata*) .............................. Front Cover

APPROACH TO
THE BORE GHAUT ............... Front Endpaper

SCLATER'S MONAL
(*Lophophorus sclateri*) .......................... Title Page

EMPEROR AKBAR HUNTING
WITH CHEETAHS ...................... 16

BLOSSOM-HEADED PARAKEET
(*Psittacula roseata*) ................................ 21

WHITE-BROWED WAGTAIL
(*Motacilla maderaspatensis*) .................. 23

WHITE WAGTAIL
(*Motacilla alba*) .................................... 24

SURPRISE APPEARANCE OF A TIGER ............ 28

BLYTH'S KINGFISHER
(*Alcedo hercules*) .................................. 33

COMMON KINGFISHER
(*Alcedo atthis*) ...................................... 34

TIGER WITH KILL ............................ 36–37

RED-MANTLED ROSEFINCH
(*Carpodacus rhodochlamys*) .................. 39

SCARLET FINCH
(*Haematospiza sipahi*) .......................... 40

TIBETAN SNOWFINCH
(*Montifringilla adamsi*) ........................ 43

SPECTACLED FINCH
(*Callacanthis burtoni*) .......................... 44

MAHRATTA WOODPECKER
(*Dendrocopos mahrattensis*) .................. 47

HEART-SPOTTED WOODPECKER
(*Hemicircus canente*) ............................ 48

SNOW PIGEON
(*Columba leuconota*) ............................ 54

RED-HEADED TROGON
(*Harpactes erythrocephalus*) .................. 57

MALABAR TROGON
(*Harpactes fasciatus*) ............................ 59

PURPLE SUNBIRD
(*Nectarinia asiatica*) .............................. 63

FIRE-TAILED SUNBIRD
(*Aethopyga ignicauda*) .......................... 64

LITTLE FORKTAIL
(*Enicurus scouleri*) ................................ 66

KHEDDA, TRAPPING ELEPHANTS ............ 71

PEREGRINE FALCON
(*Falco peregrinus babylonicus*) .............. 75

GREY TREEPIE
(*Dendrocitta formosae*) .......................... 77

RUFOUS TREEPIE
(*Dendrocitta vagabunda*) ...................... 78

CHANGEABLE HAWK EAGLE
(*Spizaetus cirrhatus*) ............................ 81

EURASIAN EAGLE OWL
(*Bubo bubo bengalensis*) ........................ 83

ORIENTAL BAY OWL
(*Phodilus badius*) .................................. 84

RUFOUS-VENTED LAUGHINGTHRUSH
(*Garrulax gularis*) ................................ 89

CHESTNUT-BELLIED ROCK THRUSH
(*Monticola rufiventris*) .......................... 91

YELLOW-THROATED LAUGHINGTHRUSH
(*Garrulax galbanus*) .............................. 93

RED-FACED LIOCICHLA
(*Liocichla phoenicea*) ............................ 95

WHITE-BREASTED WATERHEN
(*Amaurornis phoenicurus*) ...................... 99

GREEN COCHOA
(*Cochoa viridis*) .................................... 103

THE COMMON STRIPED SQUIRREL
(*Funambulus palmarum*) ...................... 104

BLACK-BELLIED TERN
(*Sterna acuticauda*) .............................. 109

| | |
|---|---|
| DESERT WHEATEAR (*Oenanthe deserti*) .......... 110 | RUFOUS-BELLIED NILTAVA (*Niltava sundara*) .......... 164 |
| MANGROVE PITTA (*Pitta megarhyncha*) .......... 115 | LONG-TAILED BROADBILL (*Psarisomus dalhousiae*) .......... 169 |
| BLUE PITTA (*Pitta cyanea*) .......... 117 | SILVER-BREASTED BROADBILL (*Serilophus lunatus*) .......... 171 |
| BRIGHT-HEADED CISTICOLA (*Cisticola exilis tytleri*) .......... 121 | SATYR TRAGOPAN (*Tragopan satyra*) .......... 173 |
| BABUR HUNTING RHINOCEROS .......... 127 | GOLDEN-FRONTED LEAFBIRD (*Chloropsis aurifrons*) .......... 177 |
| SQUIRRELS, A PEACOCK AND PEAHEN, SARUS CRANES, AND FISHES .......... 128 | TIGER ACROSS A RIVER .......... 179 |
| NICOBAR PARAKEET (*Psittacula caniceps*) .......... 131 | ORANGE-BELLIED LEAFBIRD (*Chloropsis hardwickii*) .......... 180 |
| LONG-TAILED PARAKEET (*Psittacula longicauda*) .......... 132 | JERDON'S CHLOROPSIS (*Chloropsis cochinchinensis jerdoni*) .......... 183 |
| SUNDA TEAL (*Anas gibberifrons*) .......... 135 | COMMON STARLING (*Sturnus vulgaris*) .......... 185 |
| BAER'S POCHARD (*Aythya baeri*) .......... 136 | MOTTLED WOOD OWL (*Strix ocellata*) .......... 189 |
| RED-HEADED BUNTING (*Emberiza bruniceps*) .......... 140 | GRASS OWL (*Tyto capensis*) .......... 191 |
| SILVER-EARED MESIA (*Leiothrix argentauris*) .......... 143 | SPOTTED NUTCRACKER (*Nucifraga caryocatactes*) .......... 193 |
| RED-RUMPED SWALLOW (*Hirundo daurica*) .......... 147 | ROSY MINIVET (*Pericrocotus roseus*) .......... 194 |
| PACIFIC SWALLOW (*Hirundo tahitica domicola*) .......... 148 | SHORT-BILLED MINIVET (*Pericrocotus brevirotris*) .......... 201 |
| WIRE-TAILED SWALLOW (*Hirundo smithii*) .......... 151 | WHITE-BELLIED MINIVET (*Pericrocotus erythropygius*) .......... 202 |
| STREAK-THROATED SWALLOW (*Hirundo fluvicola*) .......... 153 | SMALL MINIVET (*Pericrocotus cinnamomeus*) .......... 207 |
| YELLOW-BILLED BLUE MAGPIE (*Urocissa flavirostris*) .......... 157 | SCARLET MINIVET (*Pericrocotus flammeus*) .......... 209 |
| COMMON GREEN MAGPIE (*Cissa chinensis*) .......... 159 | FORTRESS OF BOWRIE IN RAJPUTANA .......... Back Endpaper |
| HIMALAYAN FLAMEBACK (*Dinopium shorii*) .......... 163 | BEAUTIFUL NUTHATCH (*Sitta formosa*) .......... Back Cover |

# Foreword

This book takes us back nostalgically to life in India as it existed a hundred and fifty, or even two hundred years ago. It may have been a life of discomfort – there was no electricity, no railways or motor cars, and India was more rural than it is today. But it was a slow-track, leisurely life with plenty of time for indulging in hobbies.

Our country was then ruled by the British, being the jewel in their colonial crown. Our "masters", accustomed as they were to their cold, damp, almost perpetually foggy climate with very few dangerous wild animals, must have found it very different and often very difficult, to live in an alien land and to adjust to our hot, sunny, tropical country, with pestilence to boot. But they sought solace in "taking to the hills" during the hot summer and otherwise passing their time in what seems to have been their favourite hobby and "sport": shikar.

In the 21st century many people speak in derogatory terms of shikar as a brutal, bloodthirsty way of slaughter. But before passing judgement one has to consider the conditions then existing. Our villages were tiny clearings surrounded by dense jungle teeming with dangerous wild animals. Tigers were the most notorious among these, and scarcely a day passed without a poor villager or his cattle falling prey to these marauding beasts. The ryots looked up to their *Sarkar ma-bap* and beseeched them to rid them of the brute's depredations. The *gora-log* then had to go after the cause of the hapless villagers' curse and hunt it down. Of course much shikar was done purely for sport, and huge numbers of animals and birds became game for the hunters. However, the shikaris had a code of honour and sportsmanship – never to kill a pregnant female, or one with cubs. And if their quarry did not die outright, the shikari had to follow it, often at grave risk to his life, and relieve the wounded beast from its misery. Many a shikari has met a tragic and gory end when attacked by a lurking wounded tiger.

A shikari had to pit his skill against his cunning quarry. He had to observe the animal's behaviour, its personality so to speak, in order to be successful in his venture. And invariably the hunter gradually metamorphosed into an amateur wildlife biologist and true conservationist. No example of this can be more truly cited than that of Jim Corbett, after whom our first National Park was named. Shikar as a hobby was also indulged in by the native princes (maharajas), who had exclusive forest patches in their kingdoms for this purpose. It is these very reserves which were protected and therefore retained their virgin wildlife, while the "common ground" was degraded in independent India. They form the bulk of our National Parks today.

If we look at the monumental published works on India's flora and fauna, we find that none of the authors were academically trained like today's science graduates. They were invariably from the Indian Civil Service, the Army and, later, the Railways. Outstanding among them are M.A. Wynter-Blyth (butterflies), Sir Francis Day (fishes), A. Alcock (crustaceans), T.J. Jerdon, Allan Hume, and E.C. Stuart Baker (birds), R.A. Sterndale, S.H. Prater, and T.J. Roberts (mammals). Many avid wildlifers today "hunt" with their cameras instead of a rifle. In those days there were no cameras, so they took recourse to painting. Some of the best paintings were those executed by Elizabeth and John Gould, later published as lithographs in the mid-nineteenth century. Sterndale and others are equally noted for their black-and-white renderings of mammals; so is Forbes for his *Oriental Memoirs*.

The role of the Society's Journal is no less significant. Started in 1886, it served as the mouthpiece of the shikari-cum-amateur wildlife scientist of a bygone era. Often written in a lucid, pithy style, the articles remind us of the role they played in the advancement of knowledge of our wildlife.

It was in the heyday of India's wildlife renaissance (1883) that the Bombay Natural History Society was founded. It came into being when eight men met to discuss and learn more about plants and animals. Two of them, Dr. Atmaram Pandurang Tarkhad and Dr. Sakharam Arjun, were Indians. The first few meetings were held in the Victoria and Albert Museum (now Bhau Daji Lad Museum). But, soon after in its infancy, the Society was taken under the benevolent wing of Phipson's – one of the "boxwallahs", as the merchant princes were then derogatorily called by the Civil Service and Army. Until 1948, when the last of the Phipsons left India, the Society was lodged in Phipson House.

The Society's membership grew, and it became the repository of skins of birds and animals which shikaris sent to it. With the ban on hunting today, these form a priceless collection referred to by specialists from all over the world. For example, the Society has the skin of a Pink-headed Duck, now feared extinct. Many of these shikaris and "boxwallahs" also donated books to the Society's library; these formed the nucleus of today's published collections. Among these, probably the most munificent was that of Mr. F.V. Evans of West Derby, Liverpool. Many of these books – already valuable when donated in 1925 – are extremely rare and priceless today. Some of them date back 100, 150, and even 200 years.

Very few people get a chance to read these invaluable books. So two of our intrepid members, Dr. Ashok S. Kothari and Dr. B.F. Chhapgar, have been busy for the last three years selecting the cream of lithographs and articles on Indian wildlife. The section "Gleanings" is from the Miscellaneous Notes section of our Society's Journal. The outcome of their effort is this book – a sequel to their highly popular, similar collection *Sálim Ali's India*, which came out in 1996 as a tribute to independent India's greatest ornithologist.

The joint editors have been greatly encouraged and helped by the Society's staff as well as the munificence of many sponsors.

I hope you get as much pleasure from this book as we did in compiling it.

**B.G. Deshmukh**
PRESIDENT

# Preface

The tremendous success of our earlier venture, *Sálim Ali's India* in 1996 has encouraged us to compile what may be termed as its sequel.

The British during Queen Victoria's era were fascinated by the teeming wildlife in India's jungles and the scintillating colours of its flora and fauna. While the shikaris wrote down their experiences in books, artists like Elizabeth Gould, J.C. Keulemans, Henry C. Richter, William Hart, E.C.S. Stuart Baker, Robert A. Sterndale, Joseph Hooker, James Forbes, Warwick Reynolds, William Kuhnert and others drew and painted vividly what they saw.

Our Society's library has several such books containing lithographs – many donated by its members. Special mention must be made of Mr. F.V. Evans whose donation includes the classic six volumes of John Gould's *Birds of Asia* and (with Elizabeth Gould's paintings) *Birds from the Himalaya Mountains* and, more recently, Van Ingen's vast collection of shikar and natural history books. We have culled freely from them for our present volume.

Apart from these, many shikaris-turned-naturalists contributed articles, based on their experiences, to our Society's Journal, which recently completed its hundredth volume. The section "Gleanings" contains many of these interesting episodes.

This time we have thought fit to dedicate our book to the Cheetah or Hunting Leopard. So different from the more well known leopard, the cheetah was a popular pet in the menageries of our maharajas, being used to hunt deer and small game by running them down at speeds exceeding 96 kilometres an hour. The last three of these were gunned down on the same night by an Indian "shikari" at Korea (Madhya Pradesh) as recently as 1947.

**Editors**

NOTE
We have taken care to keep editorial changes to a minimum, so as to adhere to the writer's style. Even in cases of old-fashioned spelling and punctuation we have in most cases followed the original. Only scientific names in current usage have been substituted for earlier ones.

The captions and descriptions in the margins do not pertain to the text article next to them, but to the plates facing them, or to the black and white illustrations.

# Acknowledgements

It would be impossible to thank all those whose suggestions and help have shaped this volume. We, however, wish to mention the names of those who have inspired and guided us in this endeavour. Sohrabji P. Godrej, Mrs. Vijaya Deshmukh, J.S. Serrao, Humayun Abdulali, and Professor Indumati S. Kothari are not in this world anymore but they were very keen that we take up this project and helped in many ways during the initial stages. We thank Shri B.G. Deshmukh, President of the Bombay Natural History Society, Pheroza Godrej and Dilnavaz Variava, both Vice-Presidents, J.C. Daniel, former Hon. Secretary, Dr. A.R. Rahmani, Director, Dr. Rachel Reuben, Hon. Secretary of the Society, Debi Goenka, Hon. Treasurer, and Rishad Naoroji, Chairman, Library Sub-committee for constant encouragement and moral support. Our special thanks to Dr. M.R. Almeida and Varad B. Giri for critically going through the scientific nomenclature. Pheroza Godrej, Aasheesh Pittie, and Nikin Mehta (Dallas, Texas) made extra efforts to help in collecting the necessary amount towards producing this book. Ashok Kumar Mehra (President, Rotary Club of Bombay Seacoast), Premal Parikh, Hardik Kapadia, and Devang Shah extended help whenever needed. The success of our previous book *Sálim Ali's India* helped in convincing many to provide financial support. The Palanpuri community of diamond merchants, known for its generosity, extended help by sponsoring many plates. Saroj D. Trivedi (Ahmedabad), the late Abdullabhai Carrol, 94-year-old Champaklal Shah, Messrs. Ranjan Gupta (Kolkata), and Dilnavaz Mehta (Rare Finds), all dealers in rare books, helped us in securing out-of-print books for specific articles. The late Mr. F.V. Evans donated extremely precious books to the Society's library including the famous *Birds of Asia*. Many of the plates in this book are selected from those precious tomes. A generous donation from Ravi Singh, former Executive Committee member of the Society enabled the buying of a fire-proof cabinet to store this heritage. Mr. Van Ingen's recent donation of precious books from his rich collection helped in our search for articles, especially on the wonderful wildlife of India in bygone days. We wish to thank the National Museum (New Delhi) and Victoria & Albert Museum (London) for giving us permission to print Mughal paintings from their collections.

We owe a particular debt of gratitude to the entire staff of the Society for helping in this venture. Naresh Chaturvedi, Curator, Dr. Gayatri Ugra, Publications Officer, Vibhuti Dedhia, Editorial Assistant, and Isaac Kehimkar, Public Relations Officer gave valuable suggestions. We are thankful to Tarendra Singh and Sadanand Shirsat of the library, Shabnam Shaikh of accounts, and Shalet Alva, stenotypist, who were always ready to help. We also thank Mohan Achary and Joe Rego of Infomedia India for their help in matters of printing.

Hansa Kothari supported the idea and worked patiently and silently for the book; we appreciate her help and suggestions.

There is a long list of generous sponsors without whose help this book would not have taken final shape; we are indeed indebted to them.

Finally, we are most thankful to Radhika Sabavala, General Manager of Marg Publications and the entire Marg team for undertaking this project. Our special thanks to Rivka Israel and Naju Hirani for working hard and infusing new ideas. Marg has established a permanent relationship with the Bombay Natural History Society, which will grow in years to come.

# The Founders of the Bombay Natural History Society

W.S. Millard

The Society was formed on the 15th September 1883 by, I believe, eight residents of Bombay whose names were:
Dr. D. MacDonald
Mr. E.H. Aitken
Col. C. Swinhoe
Mr. J.C. Anderson
Mr. J. Johnston
Dr. Atmaram Pandurang
Dr. G.A. Maconochie
Dr. Sakharam Arjun

In the proceedings of a Meeting of the Society held on the 22nd April 1902, (Vol. XIV, page 408), the late Mr. Aitken in referring to the approaching retirement of Dr. D. MacDonald, said, "probably few of those present knew the real origin of the Bombay Natural History Society, or had any idea that Dr. MacDonald was the *fons et origo* of the whole thing. But such was the fact. It was early in 1883 that Dr. MacDonald suggested that it would be an excellent thing to form a Society for the study of Natural History."

Mr. Aitken then mentions that six gentlemen met in the Victoria and Albert Museum [Byculla, Bombay] and constituted themselves the Bombay Natural History Society. The six names he mentions are those given above but he omits the names of Col. C. Swinhoe and Dr. Sakharam Arjun.

In the Introduction in the first number of the Journal, January 1886, it is stated that the Society was founded "by seven gentlemen interested in Natural History, who proposed to meet monthly and exchange notes, exhibit interesting specimens and otherwise encourage one another", but unfortunately it does not mention the names of the seven. The eight names given above were traced from the earliest Minute Book of the Society.

As Vol. I, No. 1, of the Journal may not be readily available to many present-day members, it may not be superfluous to quote further from the Introduction. The name of Mr. H.M. Phipson was not amongst the original founders, as he was on leave in England at the time, but he must have joined the Society very soon afterwards, as the Introduction states:–

"For several months meetings were held in the 'Victoria and Albert' Museum, but in January 1884, Mr. H.M. Phipson kindly offered the use of a room in his office in the Fort. This removal to a central situation gave an astonishing impulse to the Society. The meetings were better attended, the membership increased and collections began to be made, so that in a very short time the necessity for more ample accommodation was pressingly felt. The committee appointed to seek for suitable rooms having failed elsewhere, recommended the Society to ask Mr. Phipson to let one-half of his office premises, including the room of which they had up to this time had the gratuitous use. He consented to this and so the Society continued to hold its meetings and keep its collections at 18, Forbes Street. Its progress was so rapid however, that these premises were soon felt to be too small and ultimately the collections were removed to larger and in every way more suitable rooms at 6, Apollo Street."

It is well known that Mr. H.M. Phipson was the backbone of the Society from March 1886, – when he took over the position of Honorary Secretary from the late Mr. E.H. Aitken – to 1906, when he left India, and the success of the Society has been greatly due to his devoted labour on its behalf, and his wonderful personality, aided by other stalwart early members, such as the late

Mr. E.H. Aitken
Dr. D. MacDonald
Mr. Justice H.M. Birdwood
Mr. R.A. Sterndale
Mr. G.W. Vidal, C.S.
Mr. J.C. Anderson
Surgeon K.R. Kirtikar
Mr. W.F. Sinclair, C.S.
Rev. F. Dreckmann, S.J.
Mr. R.C. Wroughton

all of whom have passed away, but their work is evident as shown by the present flourishing condition of the Society, which has been so ably helped by so many past and present workers whose names are too numerous to mention in this brief note. Mr. H.M. Phipson is, fortunately, still with us, living in England, and continues to take a great interest in the work of the Society.

Tunbridge Wells, Kent, November 1930.
From *JBNHS*, Vol. XXXV.

**MR. MILLARD** was a link with the old founders. He joined the Society in the year 1888 and, working in close collaboration with Mr. H.M. Phipson whilst he was in India, he carried on his work as Honorary Secretary from the time of Mr. Phipson's retirement until the time came for him too to leave India – in April 1920.

**Emperor Akbar Hunting with Cheetahs**
From the *Akbar Nama* reproduced with the kind permission of
the Trustees of the Victoria & Albert Museum, London.

**Courtesy Ion Exchange (India) Ltd. – The Power Behind Water**

# Selections from Books, Journals, and Gazetteers

## Banian Tree, the Pride of Hindostan

James Forbes

**< Emperor Akbar Hunting with Cheetahs**

An illustrated copy of the *Akbar Nama* is in the collection of the Victoria & Albert Museum, London. It was prepared for the library of the Mughal Emperor Akbar (1556–1605). It bears the signatures of his son Jahangir, and a seal of great-grandson Aurangzeb. During the decline of the Mughal empire this *Akbar Nama* fell into the hands of one Ahmed Ali Khan in 1793. It was purchased by Major-General John Clark, the Commissioner of Oudh in 1896, and was acquired by the Victoria & Albert Museum from his widow. The manuscript contains 274 folios and 117 paintings. It was illustrated by 56 artists.

Akbar had a large number of Hunting Leopards in his Cheetah-Khana. This painting by Sarwan shows Akbar hunting spotted deer and blackbuck with the help of cheetahs (c. 1590). All the details, including the armour, tent walls, and animals, are rendered with painstaking care.

The Banian, or Burr tree (*Ficus bengalensis*) is deserving of our attention: from being one of the most curious and beautiful of nature's productions in that genial climate, where she sports with the greatest profusion and variety. Each tree is in itself a grove, and some of them are of an amazing size; as they are continually increasing, and, contrary to most other animal and vegetable productions, seem to be exempted from decay: for every branch from the main body throws out its own roots, at first in small tender fibres, several yards from the ground, which continually grow thicker; until, by a gradual descent they reach its surface; where striking in, they increase to a large trunk, and become a parent tree, throwing out new branches from the top. These in time suspend their roots, and, receiving nourishment from the earth, swell into trunks, and shoot forth other branches; thus continuing in a state of progression so long as the first parent of them all supplies her sustenance.

A banian tree, with many trunks, forms the most beautiful walks, vistas, and cool recesses, that can be imagined. The leaves are large, soft, and of a lively green; the fruit is a small fig, when ripe of a bright scarlet; affording sustenance to monkeys, squirrels, peacocks, and birds of various kinds, which dwell among the branches.

The Hindoos are peculiarly fond of this tree; they consider its long duration, its out-stretching arms, and over-shadowing beneficence, as emblems of the Deity, and almost pay it divine honours. The Brahmins, who thus "find a fane in every sacred grove," spend much of their time in religious solitude under the shade of the banian-tree; they plant it near the *dewals*, or Hindoo temples, improperly called Pagodas; and in those villages where there is no structure for public worship, they place an image under one of these trees, and there perform a morning and evening sacrifice.

These are the trees under which a sect of naked philosophers, called Gymnosophists, assembled in Arrian's days; and this historian of ancient Greece gives us a true picture of the modern Hindoos; "In winter the Gymnosophists enjoy the benefit of the sun's rays in the open air; and in summer, when the heat becomes excessive, they pass their time in cool and moist places, under large trees; which, according to the accounts of Nearchus, cover a circumference of five acres, and extend their branches so far, that ten thousand men may easily find shelter under them."

There are none of this magnitude at Bombay; but on the banks of the Nerbudda I have spent many delightful days with large parties, on rural excursions, under a tree supposed by some persons to be that described by Nearchus, and certainly not at all inferior to it. High floods have at various times swept away a considerable part of this extraordinary tree; but what still remains is near two thousand feet in circumference, measured round the principal stems; the over-hanging branches, not yet struck down, cover a much larger space; and under it grow a number of custard-apple, and other fruit trees. The large trunks of this single tree amount to three hundred and fifty, and the smaller ones exceed three thousand: each of these is constantly sending forth branches and hanging roots, to form other trunks, and become the parents of a future progeny.

This magnificent pavilion affords a shelter to all travellers, particularly the religious tribes of Hindoos; and is generally filled with a variety of birds, snakes, and monkeys: the latter have often diverted me with their antic tricks; especially in their parental affection to their young offspring; by teaching them to select their food, to exert themselves, in jumping from bough to bough, and then in taking more extensive leaps from tree to tree; encouraging them by caresses when timorous, and menacing, and even beating them, when refractory. Knowing by instinct the malignity of the snakes, they are most vigilant in their destruction: they seize them when asleep by the neck, and running to the nearest flat stone, grind down the head by a strong friction on the surface, frequently looking at it, and grinning at their progress. When convinced that the venomous fangs are destroyed, they toss the reptile to their young ones to play with, and seem to rejoice in the destruction of the common enemy.

On a shooting party under this tree, one of my friends killed a female monkey, and carried it to his tent; which was soon surrounded by forty or fifty of the tribe, who made a great noise, and in a menacing posture advanced towards it: on presenting his fowling-piece, they retreated, and appeared irresolute, but one, which from his age and station in the van, seemed the head of the troop, stood his ground, chattering and menacing in a furious manner; nor could any efforts less cruel than firing drive

**View of Cubbeer-Burr, the Celebrated Banyan Tree on an Island in the Nerbudda**, drawn from Nature 1778. *Oriental Memoirs*, Vol. III, by James Forbes, 1813.

him off: he at length approached the tent door; and when finding his threatenings were of no avail, he began a lamentable moaning, and by every token of grief and supplication, seemed to beg the body of the deceased: on this, it was given to him: with tender sorrow he took it up in his arms, embraced it with conjugal affection, and carried it off with a sort of triumph to his expecting comrades.

The artless behaviour of this poor animal wrought so powerfully on the sportsmen, that they resolved never more to level a gun at one of the monkey race.

The banian tree I am now describing, is called by the Hindoos Cubbeer-Burr, in memory of a favourite saint, and was much resorted to by the English gentlemen from Baroche. Putnah was then a flourishing chiefship, on the banks of the Nerbudda, about ten miles from this celebrated tree. The chief was extremely fond of field diversions, and used to encamp under it in a magnificent style; having a saloon, dining-room, drawing-room, bed chambers, bath, kitchen, and every other accommodation, all in separate tents; yet did this noble tree cover the whole; together with his carriages, horses, camels, guards, and attendants. While its spreading branches afforded shady spots for the tents of his friends, with their servants and cattle. And in the march of an army, it has been known to shelter seven thousand men.

Such is the banian tree, the pride of Hindostan, which Milton has thus discriminately and poetically introduced into his *Paradise Lost*:

Then both together went
Into the thickest wood; there soon they chose
The fig-tree. Not that tree for fruit renown'd,
But such, and at this day to Indians known
In Malabar or Decan, spreads her arms,
Branching so broad and long, that in the ground
The bended twigs take root, and daughters grow
About the mother tree; a pillar'd shade
High over-arch'd, and echoing walks between:
There oft the Indian herdsman shunning heat,
Shelters in cool, and tends his pasturing herds,
At loop-holes cut through thickest shade.

From *Oriental Memoirs*, Vol. I,
by James Forbes, 1812.

The birds, monkeys, and serpents abounding in Cubbeer-Burr are well known. The enormous bats which darken its branches frequently exceed six feet in length from the tip of each wing, and from their resemblance to that animal, are not improperly called flying-foxes. Bats of this magnitude are a kind of monster, extremely disagreeable both in smell and appearance. They must have been the harpies mentioned by Virgil.

When from the mountain-tops with hideous cry,
And clattering wings the hungry harpies fly;
They snatch the meat, defiling all they find;
And parting, leave a loathsome stench behind.

These large bats, like the rest of their species, suspend themselves by the claw, or hook on the wings with their heads downwards, when they repose or eat, in which posture they hang by thousands in the shades of Cubbeer-Burr. Archdeacon Paley remarks, that "the hook in the wing of a bat is strictly mechanical, and also a *compensating* contrivance. At the angle of its wing there is a bent claw, exactly in the form of a hook, by which the bat attaches itself to the sides of rocks, caves, and buildings, laying hold of crevices, joinings, chinks, and roughnesses. It hooks itself by this claw, remains suspended by this hold, takes its flight from this position, which operations compensate for the decrepitude of its legs and feet. Without her hook, the bat would be the most helpless of all animals. She can neither run upon her feet, nor raise herself from the ground; these inabilities are made up to her by the contrivance in her wing; and in placing a claw in that part the Creator has deviated from the analogy observed on winged animals. A singular defect required a singular substitute."

From *Oriental Memoirs*, Vol. III,
by James Forbes, 1813.

**JAMES FORBES** was an officer in the service of the East India Company at Bombay. In 1775 he accompanied the British Mission sent to support Raghoba (Raghunath Rao, the 6th Peshwa) in Gujarat and was appointed Collector of Dabhoi (near Baroda) in 1780, where he remained till 1783.

A good draughtsman and a keen observer, during his sojourn in western India Forbes filled 150 folio volumes (52,000 pages) with sketches and notes on the flora, fauna, manners, religion, and monuments of India. His lavishly illustrated *Oriental Memoirs* appeared in four volumes in 1812–13.

Forbes was very much impressed by the indigenous trees of India and particularly by India's most popular tree, the Banyan tree (*Ficus bengalensis*). He visited the celebrated "Kabir Vad" near Broach and described the tree which by its gigantic spread had taken the form of a small forest on a tiny island on the Narmada.

**REV. REGINALD HEBER**, in his *Narrative of a Journey…* (Vol. II, 1844) writes of the same tree: "Another curiosity in the neighbourhood of Broach is the celebrated *bur* or banian tree, called *Kuveer Bur*, from a saint who is said to have planted it. It stands on, and entirely covers an island of the Nerbudda, about twelve miles above Broach. Of this tree…which the natives tell us boasted a shade sufficiently broad to shelter ten thousand horse, a considerable part has been washed away with the soil on which it stood, within these few years by the freshets of the river; but enough remains, as I was assured, to make it one of the noblest groves in the world, and well worthy of all the admiration which it has received."

The late **SOHRABJI P. GODREJ**, former Vice-President of the BNHS and a great tree lover himself, always insisted that the Banyan tree should be named "The Tree of India".

**Blossom-headed Parakeet >**
*Psittacula roseata*
Head and cheeks rosy pink, rear crown pale lilac-blue, back green, small yellow bill, tail has pale yellow tip. Female has paler greyish blue head, yellowish green hind collar. Prefers well wooded areas, open forest and cultivation in forest clearings. Resident, mainly NE Indian hills (Arunachal Pradesh, Assam, Meghalaya, Sikkim, Tripura, West Bengal) and Bangladesh.

ARTICLE AND ILLUSTRATION COURTESY
SREEKANT S. MEHTA, FAIRLANDS, SALEM, TAMIL NADU

**Blossom-headed Parakeet** *Psittacula roseata* Biswas
*Birds of Asia*, Vol. V, Parts XXV–XXX, by John Gould, 1873–1877. Painted by John Gould & Henry C. Richter.

Courtesy Central Bank of India

# The Manpoora Tiger
## About a Tiger Hunt in Rajpootanah

"Royal Tiger"

After my arrival at Beawr in Rajpootanah, the first point about which I made enquiry was, of course, "are there any tigers to be met with here?" I received the gratifying information, "that there were, but," added my informant, "it is impossible to get at them on account of the hilly country." The fact of tigers being in the neighbourhood was quite sufficient for me; and I felt perfectly well satisfied that I should, sooner or later, come in contact with them, in spite of hills, or any other obstacles.

So soon as the warm weather had begun to set in, so soon my thoughts turned towards my friends the tigers. The country was desperately hilly it is true; but previous to devising any other plans, I determined on ascertaining how far it might be practicable to make use of elephants. They were, accordingly, ordered out; and after going over several most break-neck-like-looking places, I was compelled to give it up as a bad job; and, I confess, that I felt no great sorrow at not having stumbled on a tiger, as a charge home from one of them, might have dashed me, elephants and all, into eternity, and no mistake. I was now fully convinced, that unless I could catch a fellow napping in the plains, I might leave elephants out of the question; but this was never the case, – the tiger, as if aware of his perfect security in the hills, kept to them during the day, and quit them only by night. I became fully satisfied that night work was my only chance; and I was the more convinced on this point, after a morning's tour round the Kalingur Lake (this lake is an artificial one, and a beautiful piece of water it is), when I observed some very fresh tracks of tigers, and on enquiry among the villagers, I was informed, that I could meet with *janwars* in abundance any night round Kalingur. From this moment I resolved on nocturnal excursions: but how to proceed I was at a loss; for there was not a tree on which I might erect a *mechan*, nor a blade of grass, into which I might conceal myself round the lake, – the ground about it being as bare as the table I am writing on; – so what was to be done? I must either watch on the bare ground or get no tiger: the choice left me was not of a very tempting nature; and I hardly relished the idea of coming in contact with a tiger on an equal footing with him, and that particularly at night. I paused for a moment, and then as if awaking from a dream, the resolution of taking my chance was formed. I thought I could depend on my shot, and if my nerves would but remain firm, the result I did not much fear.

The lake already named, was at a distance of six miles from my bungalow, and the road to it lay through very heavy *jungle*. My departure from home was always taken so as to admit of my reaching the spot a little after sunset, where I was obliged to remain till daybreak the following morning. My companion on these excursions was a young *Mussulman* lad, in the capacity of *kitmetgar*, who had charge of the ammunition &c., and whom I armed with a hog spear.

**White-browed Wagtail >**
*Motacilla maderaspatensis*
A large black and white wagtail of nearly the colour pattern of the Magpie Robin with a conspicuous white streak over the eye from nostril to behind the ear. Head, upper breast, and entire plumage and wings black; a broad tapering white band running the whole length of the folded wing. The forehead is broadly white. Female has the black portions duller and browner. Commonly found singly or in small family parties at water's edge, walking about, searching for insects, the long tail incessantly wagging. Makes a loud *chiz-zit* call during flight; the song is a clear high-pitched jumble of loud, pleasant whistling notes reminiscent of Magpie Robin. Resident, occurs more or less throughout India and Sri Lanka.

White-browed Wagtail (Large Pied Wagtail) *Motacilla maderaspatensis* Gmelin
*Birds of Asia*, Vol. I, Parts I–VI, by John Gould, 1850–54. Painted by John Gould & Henry C. Richter.

Courtesy London Star Diamond Company (India) Pvt. Ltd.

White Wagtail (Pied Wagtail) *Motacilla alba* Linnaeus
*Birds of Asia*, Vol. III, Parts XIII–XVIII, by John Gould, 1861–66. Painted by John Gould & Henry C. Richter.

Courtesy Gujarat Ambuja Cements Ltd.

In the month of April, and after a parching day of hot winds, I made my first excursion; and my position was thus taken up:– my front facing the hills, and my back to the water, by which plan I was secure from surprise in the rear, and in case of an "untoward event," a rush into the water was to be my retreat. I confess that I did not feel particularly comfortable on this my first visit, and I hailed the morn with no small degree of delight. This night proved a blank, that is, no tiger. Of deer and hog, I saw many; but they were permitted to pass me unmolested.

A few nights after this, I was again at the spot, and the only difference from the former occasion, was, in myself, – I felt a little more at home. The hour for the moon's rising was ten; I was, therefore, taking it leisurely with my gun lying by my side, not expecting to have occasion for its immediate use; in this, however, I was most agreeably disappointed.

A little after dusk, as I was taking a glance around me, I plainly perceived some large animal, rather on one side of me, and moving towards the water; reaching which, he stopped – to take a drink, I fancy. Here was a time! – what words can describe my feelings! – and who can imagine them; who has not felt them? – A deadly silence prevailed, – not even a whisper passed between my servant and me. The faithful "Manton" was now firmly grasped, – the finger on the trigger ready to deliver the deadly fire. Worlds, at this moment, I might have given for a change of situation, but, shortly after, I would not have taken worlds for my place! Many seconds were not allowed me for reflection before I was called upon to perform a more active part; of course, I imagined, that which I saw to be nothing less than a tiger, but it might have been a samber or other kind of large deer: I, nevertheless, prepared for the worst.

After the animal, as I imagine, had refreshed himself with a drink of water, he appeared to be moving in my direction; but as the evening had considerably advanced I could not distinguish clearly; a few seconds, however, convinced me that he was. Twice the gun was brought up to the position for firing, and on each occasion, I fancied I heard a whisper, "not yet," "not yet." I took the advice and waited. A very short interval after, I observed the animal changing his direction; in fact, of this I was pretty certain, for what appeared hitherto as a round "black ball," now resembled a *long black band*. For the third and last time, (the object was now within a dozen yards from me, – I speak from a certainty, as the spot was measured the following morning) the gun was brought up; and, though I assert it myself, without the shake of a muscle. A steady aim was taken at the centre of the black band moving before me, and the fire delivered.

For the first time now, since first seeing the animal, the silence that prevailed, was broken by my servant observing, "*burra bag sahib*." I asked him how the devil he could tell, for I could not make out what it was. He observed, that from the flash of the pan, (I was using a "flint gun" then) he had plainly seen the tiger; and to my no small delight so it turned out to be; for after the rising of the moon, I observed my friend pinned to the ground on the very spot he was standing.

On the following morning the fallen foe was secured on the back of a pad elephant, and paraded round to the residents of the station. This good fortune, strengthened my confidence much, and my nocturnal excursions were consequently repeated, as often sometimes as three times a week, and my success not to be complained of. By this mode, a variety of *janwars* were killed by me, and of one, the famous Manpoora white tiger, I must give you an account.

Know then that in the neighbourhood of Manpoora, a village situated in a small valley, surrounded with hills and thick jungle, dwelt in solitary grandeur a monster of a tiger, who had become as well known as the village itself, and who had, for several

< **White Wagtail**
*Motacilla alba*

Small, dainty bird of black, white, and grey plumage. Winter plumage: forehead, anterior portion of crown, sides of head, neck, and lower plumage white, except a crescentic black band on the breast; remainder of head and nape black; upper plumage grey; wings broadly margined with white; tail black except two outer pairs of feathers which are white. In summer the chin, throat, and the upper breast are black. Female similar to male but duller. They walk swiftly on the ground, usually in parties, incessantly wagging their long tails up and down; partial to the neighbourhood of water, wading in shallow portions of it. A fairly common winter visitor to the hills and plains. Six races, viz. *M.a. dukhunensis, M.a. personata, M.a. alboides, M.a. leucopsis, M.a. ocularis, M.a. baicalensis* concern us. They arrive about September/October, departing March/April.

years past, been permitted to remain undisturbed, in consequence of his having baffled every effort of many and many a party. He, therefore, continued his depredations with perfect impunity, and became the terror of the inhabitants for miles around.

Near to the village above described, runs a beautiful little hill stream, shallow, but as clear as crystal; this struck me as being a likely spot for the promenades of this monarch, and the villagers were of my opinion also; I lost no time therefore in selecting a convenient place for myself, and hoped for a speedy interview.

For three or four nights I tried my fortune on the banks not of "Allan Water," but of "Manpoora Water," without any sort of success, and the only satisfaction I had for my trouble, was, in the morning to find, that the tiger had been prowling about either above, or below me. To watch for him, therefore, any longer after this manner, I considered useless, and accordingly gave it up, though I was resolved on not allowing the first favourable opportunity to escape me. From this time until about a period of three weeks nothing transpired worthy of being chronicled, and I had nearly given up to despair: with a view however of guarding against the consequences of low spirits, I accepted an invite to a fancy ball at Nusseerabad, a station, distant thirty miles from this out-post.

On the afternoon of the 21st of May, well do I remember it, between the hours of four and five, as I had got on my horse to ride in for this said party, my attention was arrested by the appearance of three villagers running into my compound, as if "Old Nick" himself had been at their heels, and with their hands up beckoning me to stop.

On approaching me, all that could be stammered out was, "*Manpoora ka nar aio!*" which being Anglicized, means, the "Manpoora tiger has made his appearance!" (In this part of the country a tiger is called *nar*.) From the exertion of *running*, I could

**The Tiger Tracked Down**. Drawn by George Trigger. *The Bengal Sporting Magazine*, Vol. XV, No. 89, July 1840.

not obtain for the present any further information from these villagers, but on their coming a little more to themselves, they informed me, that a cow had just been killed, and if I watched that night, the tiger would for a certainty be mine. This was sufficient; quadrilles, waltzes and country dances, were no longer thought of, agreeable partners quite forgotten, the pumps changed for the gun, and the ball room for the valley of Manpoora.

Yet the last rays of feeling, and life must depart,
Ere the scene of that valley shall fade from my heart.

Much time was not occupied in making my preparations for this unexpected expedition, and on arriving at the village, I was escorted by several villagers to the spot where the murder had been committed, and I was informed by the two individuals who had remained up in trees, that the tiger had retreated into the hills, after partaking a copious draft of pure blood. I found this to be the case as I saw the fresh kill, perfectly untouched, save the works of the tiger's claws and tusks.

I was not at all sorry to observe this, nor in the least disheartened, by the disappearance of the tiger, as I felt perfectly convinced that he would return sometime during the night for his supper, and at which, I determined to await him, and give him the warmest reception I could.

The spot all around was abominably jungly, and carried a desperate appearance about it, and not at all tempting for a night's abode. I almost looked upon it as madness to attempt it, but when I recollected the days of anxiety this tiger had caused me, and that I might not again soon meet with such an opportunity, I could not give up the idea of losing it, and therefore decided on having an interview with him if possible.

The carcase of the cow was moved by my directions to a little clearer spot, and close to the "*tail*" I formed a slight ambuscade of thorns to conceal me merely from view, and not for the sake of any protection, and when my arrangements were completed, which was after sunset, I desired the villagers to retire and to keep on the alert, in case they were called. One from among them expressed a wish to remain with me, to which I consented, and with my factotum, this made my number three – I had however to regret having allowed the villager to form one of the trio, as will be seen by what follows:

Towards dusk I took up my position, placing myself in front close up to the tail of the cow, and the two natives in my rear, and back to back, by which plan a look out on all sides was effected. The night setting in, on this occasion, I shall never forget: the darkness was horrible, and conveyed a sense of horror which, my pen cannot describe; my feelings were not the most enviable, and I could not but reflect on the folly of my conduct, and I could not answer to my conscience, what had possessed me to place myself in such a situation. These are acts designated by my friends "the mad practices of former days."

Hour after hour passed on without producing "good," or "evil," but the abominable suspense was dreadful; midnight was passing fast, and my hopes began to fade; this tranquil state was not however to last much longer, and the affair was soon to be brought to a close. Between midnight and one o'clock, a distant rustling among the bushes was distinctly heard, and by degrees it became plainer and plainer; there was no mistaking now the approach of an enemy; well do I remember the sensation on the occasion, and the thoughts that flew over one after another in quick succession.

I no longer heard the noise that had so lately occupied my attention, and I began

**Springing up to Attack a Langur**. Drawn by Warwick Reynolds. *Dwellers in the Jungle* by Lieut.-Col. Gordon Casserly, 1925.

Surprise Appearance of a Tiger
*Oriental Field Sports*, Vol. I, by Thomas Williamson, 1808. Drawn by Thomas Williamson & Samuel Howitt.

Courtesy Sunil & Sharmila Dharod, Chris & Pooja, Dallas, Texas, USA

to congratulate myself on the coast being free, when to my dismay, I declare it, I saw the beast standing close to the head of the dead cow; even at this distant period my blood freezes at the thought, and fortunate do I consider myself to have escaped as well as I did. Hardly a second elapsed after my observing the beast, before I heard him, and that distinctly, eating away at the carcase; this was rather nervous work, but it was nevertheless "now," or "never;" up came the gun, and as quickly followed the shot, but the animal did not drop; the roars he gave are fresh in my ears, and the pace at which he rushed by me, can never be forgotten. Firing a second shot was out of the question, but had it been possible, I should have been most effectually prevented by the villager, who in his fright had seized me by the arm, and in so firm a manner, that it required some exertion to extricate myself from his hold.

After the animal had passed me a short distance, I heard him fall over, and the groans which followed, convinced me that my friend whoever he was, was done, and after I had allowed sufficient time to guard, against all accidents, I desired my two attendants to hail the village. During this interval I made the attempt to reload my gun, but that I found to be an impossibility; my hand refused to perform the office, and a nervous fit had completely got hold of me; perhaps it is as well that these symptoms had not shewn themselves in the first instance.

The villagers having now assembled with torches &c. &c. we commenced a search, but strange to say nothing was to be found. I almost fancied that the whole transaction must have been a dream, though I gave directions for a second search, which was begun in right earnest. The villagers believing, that I had missed my aim, became bolder, and searched more minutely when I had the gratification to hear one of them call out "*burra nar.*" We all ran towards him, and, correct enough, there was the tiger, and he, the very identical one I had been so anxious about. I gave a hearty hurrah, in which I was joined by the motley group now assembled. I found the ball had gone clean through the centre of the stomach, and it was a matter of surprise that he had strength to proceed the distance he did.

Having now related the particulars of some of my nocturnal excursions I shall bring this epistle to a close.

From *Bengal Sporting Magazine*,
Vol. IV, 1836.

IN FOND MEMORY OF SUBASH L. CHULANI, FROM SATISH L. CHULLANI

# Baug! Baug! Baug!

Sir John Day

Matters had been thus judiciously arranged: tents were sent off yesterday, and an encampment formed within a mile and a half of the jungle which was to be the scene of our operations; and in this jungle the thickets of long rank grass and reeds are in many places fifteen feet high. At one o'clock this morning thirty elephants, with the servants, and refreshments of all kinds, were dispatched; at two we all followed in fly-palanquins; at a quarter after four we reached the encampment, and having rested near two hours, we mounted our elephants, and proceeded to the jungle.

In our way we met with game of all kinds: hares, antelopes, hog-deer, wild boars, and wild buffaloes; but nothing could divert our attention from the fiercer and more glorious game.

At the grey of the dawn we formed a line of great extent, and entered a small detached jungle. My elephant (sorely against my grain; but there was no remedy, for my driver was a keen sportsman, and he and I spoke no common language) passed through the centre, but happily no tiger had at that hour nestled there. I saw, however, as I passed through it, the bed of one, in which there were an half-devoured bullock and two human skulls; with an heap of bones, some bleached, and some still red with gore.

We had not proceeded five hundred yards beyond the jungle, when we heard a general cry on our left of Baug, baug, baug! On hearing this exclamation of Tiger! we wheeled; and, forming the line anew, entered the great jungle, when the spot where a single tiger lay having been pointed, on the discharge of the first gun a scene presented itself confessed by all the experienced tiger hunters present to be the finest they had ever seen. Five full-grown royal tigers sprung together from the same spot, where they had sat in bloody congress. They ran diversely; but running heavily, they all couched again in new covers within the same jungle. We followed, having formed the line into a crescent, so as to embrace either extremity of the jungle: in the centre were the houdar (or state) elephants, with the marksmen, and the ladies, to comfort and encourage them. In one Mr. Zoffani with Mrs. Ramus, in the other Mr. Ramus with Lady Day, led the attack; my brother and I supported them; and we were followed by Major Bateman, Mr. Crispe, Mr. Longcraft, and Mr. Van Europe, a Dutch gentleman.

These gentlemen had each an elephant to himself. When we had slowly and warily approached the spot where the first tiger lay, he moved not until we were just upon him; when, with a roar that resembled thunder, he rushed upon us. The elephants wheeled off at once; and shuffled off. They returned, however, after a flight of about fifty yards, and again approaching the spot where the tiger had lodged himself, towards the skirts of the jungle, he once more rushed forth, and springing at the side of an elephant upon which three of the natives were mounted, at one stroke tore a portion of the pad from under them; and one of the riders, panic struck, fell off. The tiger, however, seeing his enemies in force, returned, slow and indignant, into his shelter; where a heavy and well directed fire was poured in by the principal marksmen; when, pushing in, we saw him in the struggle of death, and growling and foaming he expired.

We then proceeded to seek the others, having first distinguished the spot by pitching a tall spear, and tying to the end of it the muslin of a turban. We roused the other three, in close succession, and, with little variation of circumstances, killed them all; the oldest, and most ferocious of the family, had, however, early in the conflict, very sensibly quitted the scene of action, and escaped to another part of the country.

While the fate of the last and largest was depending, more shots were fired than in the three other attacks; he escaped four severe assaults, and taking post in different parts of the jungle, rushed upon us at each wound he received with a kindled rage, and as often put the whole line to flight. In his last pursuit he singled out the elephant upon which Lady Day was; and was at its tail, with jaws distended, and in the act of rising upon his hind paws to fasten on her, when fortunately she cleared the jungle; and a general discharge from the hunters having forced him to give up the chase, he returned to his shelter.

The chase being over, we returned in triumph to our encampment, and were followed by the spoils of the morning, and by an accumulating multitude of the peasants from the circumjacent villages, who pressed round an open tent in which we sat at breakfast, with gratulations, blessings, and thanksgiving. The four tigers were laid in front; the natives viewed them with terror, and some with tears.

An old woman, looking earnestly at the largest tiger, and pointing at times to his tusks, and at times lifting his fore-paws, and viewing his talons, her furrows bathed in tears, in broken and moaning tones narrated [her sad story]. She was widowed and childless; she owed both her misfortunes to the tigers of that jungle, and most probably to those which then lay dead before her; for they, it was believed, had recently carried off her husband and her two sons grown up to manhood. In the phrenzy of her grief she alternately described her loss to the crowd, and in a wild scream demanded her husband and her children from the tigers; indeed it was a piteous spectacle!

The site of our encampment was well chosen; it was a small sloping lawn, the verdure fresh, and skirted on three sides with trees; the fourth bounded by the deep

"**On the Watch**". Drawn by Robert Armitage Sterndale. *Denizens of the Jungles*, 1886. "R.A. Sterndale was an army man in India. He was one of the first Editors of the *JBNHS* in 1886. Later he became the Governor of St. Helena." (Communication from the late J.S. Serrao, Ornithologist, BNHS.)

bed of a torrent-river. At proper distances on this lawn, there were five large and commodious tents, pitched in a semicircle: that in which we all assembled, and passed the sultry part of the day, was carpeted, and by means of the tattees of aromatic grass, continually watered, kept at a temperature pretty near to that of an April day in England. Here we had a luxurious cold dinner, with a variety of excellent wines, and other liquors, well cooled; and while we dined the French-horns and clarionets played marches, hunting-pieces descriptive of the death of the game, and other slow movements; the tigers still lying in front, and the people still assembled, but retired to a greater distance; where they anxiously waited the signal for skinning and cutting up the slain; for with them the fat of a tiger is a panacea; the tongue dried and pulverized a sovereign specific in nervous cases, and every part applicable to some use; even the whiskers they deem a deadly poison, and most anxiously, but secretly, seek them, as the means, in drink, of certain destruction to an enemy.

As my share of the spoil, I have reserved one of the talons of the tiger which pursued Lady Day, and intend to have it set in gold, with a swivel and fillet, ornamented with diamonds, and filling it with attar of roses, I shall sometime hence surprise her with it, and insist upon her giving it a place among the trinkets of her watch, as a trophy; the "spolia opima," torn from the body of an enemy slain in battle. I have reserved also a skin for you; which shall when cured be sent to you; and I shall hope to see it, ere many years elapse, an hammer-cloth to an handsome chariot of yours in the streets of London.

From *Oriental Memoirs*,
Vol. IV, by James Forbes, 1813.

The tiger hunt narrated here occurred upon the banks of the Ganges, near Chinsura in Bengal in April 1734. James Forbes says, "I have occasionally joined the European parties in their hunts, as particularly mentioned in the wilds of Turcaseer. The forests on the confines of Bhadrapur are equally wild and infested with beasts of prey. As I can offer nothing so interesting on this subject as a description of a tiger-hunt in Bengal, the subject of a letter from Sir John Day to Sir William Jones, which I have had for many years in my possession, I shall not apologise for inserting so highly-finished a picture of this royal sport; which was given to me by a very intimate friend of the writer, and has not to my knowledge appeared in print."

**Blyth's Kingfisher >**
*Alcedo hercules*

A medium-sized, dark blue and orange, forest-dwelling kingfisher with a large black bill, mouth blood-red, and legs and feet coral-red. Superficially very much like Common Kingfisher but larger. Sexes alike. It is rare, found singly, and haunts deeply shaded fast-flowing forest streams. It is shy and difficult to observe. Perches inconspicuously in low bushes overhanging water to watch for fish and aquatic insects. Loud shrill call uttered while in flight. Nest: a horizontal tunnel dug into the steep bank of a forest stream or ravine, nearly 8 cm in diameter and usually between 45 and 70 cm long, ending in a widened egg-chamber. Resident, mainly NE India (Arunachal Pradesh, Assam, Manipur, Nagaland, Sikkim), Bhutan, Bangladesh, Myanmar, North Vietnam, etc. Globally threatened.

ARTICLE AND ILLUSTRATION COURTESY
PARAM PREET SINGH, KAROL BAGH, NEW DELHI

Blyth's Kingfisher (Great Blue Kingfisher) *Alcedo hercules* Laubmann
*JBNHS*, Vol. X, No. 4, 1897. Painted by E.C. Stuart Baker.

Courtesy Sudhir Seth, Avocent India

**Common Kingfisher** *Alcedo atthis* (Linnaeus)
*Birds of Asia*, Vol. III, Parts XIII–XVIII, by John Gould, 1861–66. Painted by John Gould & Henry C. Richter.

In Memory of Dhirajlal J. Parekh & Kanchanben P. Choksi, from Sarla & Sevanti Parekh, Mumbai

# Gond Fable of Singbaba

**< Common Kingfisher**
*Alcedo atthis*

A small, brilliant turquoise-blue and orange kingfisher, with short stumpy tail and long, straight, pointed bill. Sexes alike. Found singly by stream or tank, canal or flooded ditch perched on an overhanging branch scanning the water, or flying very swiftly low over the surface of water uttering a sharp *chichee, chichee*. While sitting on a branch it bobs its head, turning it from side to side and jerks its stub-tail to the accompaniment of a subdued click. Food: small fish, tadpoles, and aquatic insects. Nest: a horizontal tunnel dug into the earth-bank of a stream, 50 cm to a metre in length, ending in widened egg-chamber. Distribution: throughout the Indian Union, Nepal, Bhutan, Bangladesh, Pakistan, Myanmar, Thailand, Malay Peninsula and islands, and the Philippines.

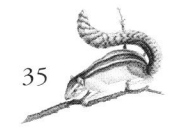

It is interesting to note that the leading idea of Mr. Rudyard Kipling's fascinating "Jungle Book" of which the scene is laid in Seoni appears to be taken from the translation of a Gond fable given in Sterndale's "Seoni", though of course stories of children being brought up by she-wolves have been reported from various parts of India.

In view of the interest attaching to the fable it may be reproduced in full here.

"The Song of Sandsumjee"

Sandsumjee married six wives, but had no heir, so he married a seventh and departed on a journey; during his absence, after his relatives had sacrificed to a god, she bore a son, Singbaba. The "small wife was sleeping, the other six were there;" so they took the babe and threw it into the buffalo's stable, placing a puppy by her side, and said "Lo! a puppy is born."

But the buffaloes took care of Singbaba and poured milk into his mouth.

When the six wives went to look for him, they found Singbaba playing.

Thence they took him and threw him to the cows, but the cows said, "Let no one hurt him," and poured milk into his mouth. So when the six wives went to look again whether he was alive or dead, lo! Singbaba was playing.

Thence they took him and threw him into a well, but on the third day when they went to enquire, they found Singbaba still playing. So they took him and threw him on the tiger's path as the tigers were coming, and they heard his cries as they left him. But the tigress felt compassion, and said, "It is my child," so she took him to her den, and having weaned her cubs fed Singbaba with milk, and so he grew up with the cubs. To her one day Singbaba said, "I am naked; I want clothes." So the tigress went and sat by the market road till muslin and cloth makers came along; on seeing her run at them they dropped their bundles and fled, which she took up and brought to Singbaba, who clothed himself and kissed her feet.

Another day he said, "Give me a bow." She again went and waited till a sepoy armed with a bow passed by. She roared and rushed at him, on which he dropped the bow and fled, and she picked it up and brought it to Singbaba who shot birds with it for his little tiger brothers.

In the meantime Sandsumjee returned home and said: "Is any one inspired? Has God entered into any one? If so, let him arise."

Then Singbaba received inspiration, and accompanied by his big and little brothers went. In the midst of the assembly was a Brahman. Him Singbaba required to get up; he refused, whereupon the big brother (tiger) got angry and did eat him up. All asked Singbaba "Who are you?"

## Tiger with Kill (Nilgai) in the Scrub Jungles of Central India

*Harmsworth Natural History: A Complete Survey of the Animal Kingdom,* Vol. I, 1910. Painted by William Kuhnert.

R.A. Sterndale writes in his *Natural History of Mammalia*: "The distinction between the Central Asian and Indian Tiger is unmistakable. The coat of the Indian animal is smooth, short hair; that of the Northern one of a deep furry pelage, of a much richer appearance. Most sportsmen recognize the difference between the stout, thick-set Tiger of the hilly country and the long bodied, lankier animal of the grass jungles and plains. Tigers are divided into (1) The *Lodhia bagh*, or the Game-killing tiger, retired in his habits, living chiefly among the hills, and retreating readily from man; (2) The *Ontia Bagh*, or Cattle-lifter, usually an old and heavier animal, very fleshy, and indisposed to severe exertion; and (3) *Adam-Khor*, or Man-eater.

**Courtesy HSBC
The Hongkong and Shanghai Banking Corporation Limited**

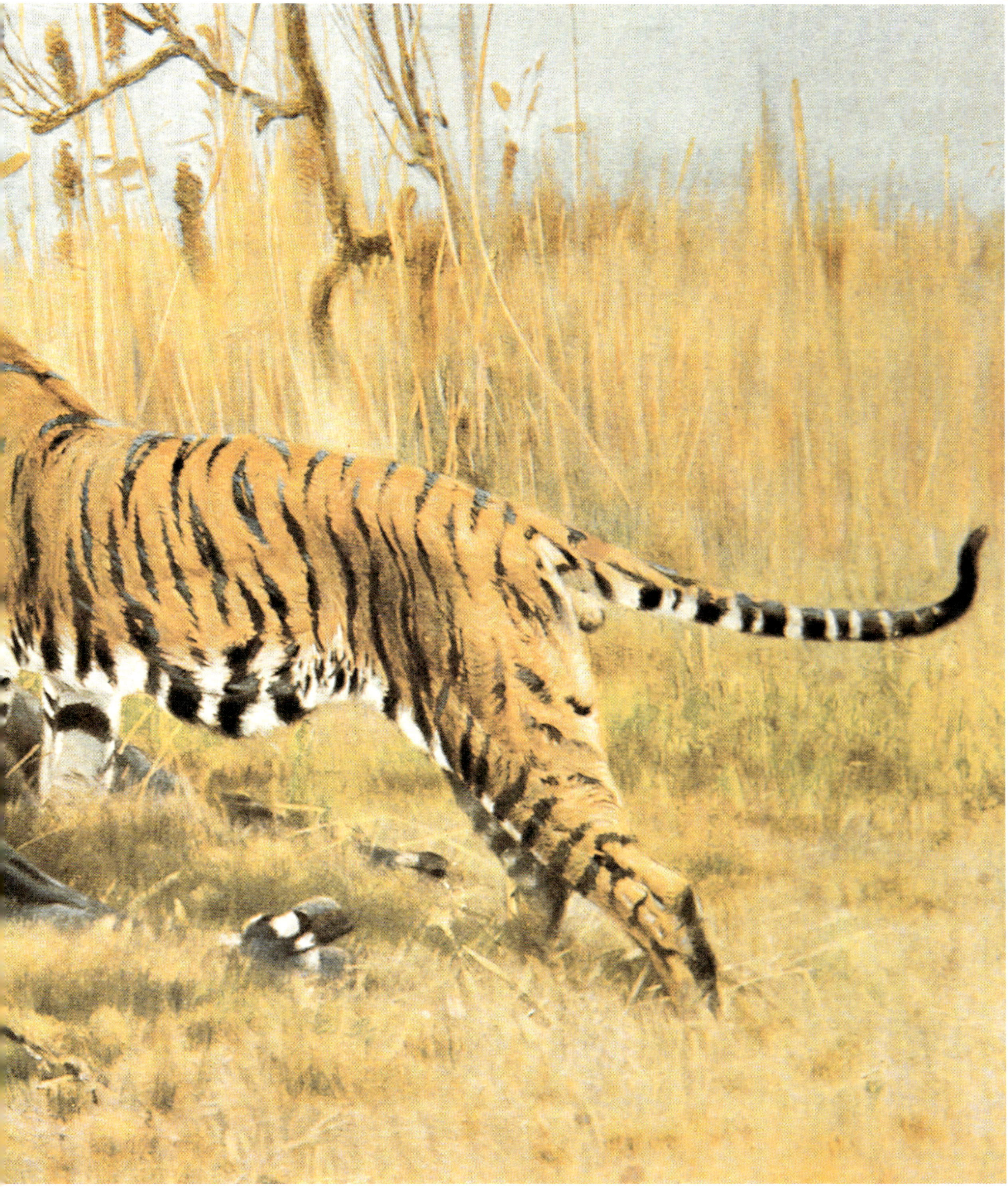

"Ask the buffaloes," he replied, telling his little brother to go and call his mother. She came, and the three species were assembled before the people. "Question them," said Singbaba. So they asked, "Who is he?" First the buffaloes answered, "Sandsumjee's son", and they told his history.

Then the cows told how he stayed with them two days and then was thrown into the well; from thence they knew not where he went.

"Ask my mother," said Singbaba.

So the tigress told how she weaned her cubs and nourished him, on which all embraced her feet and established her as a god, giving her the six wicked wives. So Singbaba became illustrious, and the tigress was worshipped.

*Sandsumjee Babana'id saka and,*
Of Sandsumjee Baba this song is,
*Bhirri bans bhirrita saka and.*
of Bhirri bamboo jungle Bhirri this song is.

From *Central Provinces District Gazetteers, Seoni District*, edited by R.V. Russel, I.C.S., 1907.

**Red-mantled Rosefinch >**
*Carpodacus rhodochlamys*
Male, crown and band behind eye reddish-brown with broad pale-pink band above eye. Back pale brown with a pinkish tinge and dark brown streaks. Lower back pink. Wings and tail rosy-brown. Below, throat and sides of head pale pink. Rest of under-parts rosy-red. Female, ashy brown above with darker streaks; whitish below, streaked with dark brown. Keeps in pairs in breeding season, in small parties in winter. Forages chiefly on ground, also in low bushes. Like other rosefinches has habit of raising feathers of crown, then looking as if crested. Resident, Pakistan from northern Baluchistan north to Chitral, thence east through Gilgit, Astor, Baltistan, Ladakh, Lahaul, Spiti, Garhwal, Kumaun (specimen in British Museum). Breeds between 2,700 and c. 3,800 m in juniper, rose bushes, and other shrubs in drier mountain forest; winters lower down in olive trees, thorn scrub, orchards, and gardens.

COURTESY NALINI & MAHENDRA P. SHAH, SHEETAL & DEVANG M. SHAH, MAAHIR & VEER, VILE PARLE (W), MUMBAI

Red-mantled Rosefinch (Blyth's Rosefinch) *Carpodacus rhodochlamys* Blyth
*Birds of Asia*, Vol. I, Parts I–VI, by John Gould, 1850–54. Painted by John Gould & Henry C. Richter.

Courtesy Ranjanben & Nikin Mehta, Mamta & Zubin Mehta,
Ami Mehta & Shaan, Dallas Diamonds Corporation, Dallas, Texas, USA

Scarlet Finch *Haematospiza sipahi* (Hodgson)
*Birds of Asia*, Vol. I, Parts I–VI, by John Gould, 1850–54. Painted by John Gould & Henry C. Richter.

Courtesy Dr. Reddy's Laboratories Limited

# The Oppression of Man

S. Eardley-Wilmot

**< Scarlet Finch**
*Haematospiza sipahi*
A stocky, stout-billed, short-tailed finch. Male brilliant scarlet with the wings and tail dusky brown. Bill yellow; legs brown. Female dusky brown; lower back bright yellow; beneath pale olive-yellow with dusky crescentic marks. Has strong and dipping, typical finch-like flight with rapid wing beats. Food: seeds, berries, flower buds, insects. Call: a pleasant soft high-pitched *too-ee*. Resident, uncommon, subject to altitudinal movements. Himalayas in N. Uttar Pradesh and from Nepal east to Arunachal Pradesh, NE India. Recorded in summer at 2,300–3,000 m. In winter down in the foothills and duars, open pine and oak forest, and tropical jungle. Scattered flocks.

As time went on the solitude which at first had been so marked a feature in the tiger's domain was broken by the ever-increasing number of human beings who found occupation or amusement therein. Formerly, only herds of cattle and their attendants roamed the forest, paying no attention to the jungle-folk, and hardly noticed by them; later on came those who felled timber and cut bamboos, their camps were numerous all over the area; and last of all came hunting parties of varying size, from the solitary sportsman who wandered afoot amongst the wild animals, to the large company, well organized to slay, who boasted of the number of their victims, and were proud of their stud of elephants and of their army of trackers and huntsmen. Not only peace but safety had departed, for though the graziers might not tell of the tiger's whereabouts, there were others, cartmen, sawyers and carpenters, who for the sake of a small reward, indeed often as a remedy for their own fears, would report all that they saw or heard to those who were able to make use of the information.

The tiger, now well experienced and cautious, gifted moreover with a most intimate acquaintance with the forest, yet found difficulty in evading all of these human beings, and their repeated invasions so seriously reduced the head of game in his hunting grounds, and forced the remainder to be so constantly on the alert that he lived in a perpetual state of anxiety, and was often put to great trouble before he could obtain a meal. Even when he had succeeded in capturing his prey he feared to return to the kill lest during his absence some ambush should have been laid; so that, unless he could drag his victim close to some water supply, he derived but one day's food from even the largest animal. For tigers must drink after a heavy meal of flesh, and particularly in the hot weather when hunting parties were abroad, he suffered torments if forced to remain thirsty for many hours.

He dreaded the approach of human beings and the loud reports of the weapons they carried; and so while eager to slink away if this were possible, yet, if by chance his retreat was cut off, his natural courage asserted itself, and was indeed fortified by his hatred of his persecutors. He had seen others fall victims to the dangers he had so far escaped; the stag with mortal wound rushing blindly through the forest only to fall dead when breasting the stony slopes; or the panther lying harmless after hours of agony; he had followed the trail of others, doomed to a lingering death but for his swift interference; and the increasing difficulties of his existence rendered him more cautious and also more morose. For to live always in fear of death results in a change of habits and characteristics, and induces a strain of unaccustomed cruelty. He had been driven by gangs of beaters, and had learned that the less risk lay in escaping through the advancing line; for, though there might be guns there, yet, in the confusion of his onslaught, these had hitherto been ineffective; while the very caution necessary to steal away through the hidden sportsmen in front afforded to these an easy shot from their posts of vantage. He had been fired at from "machans," and now was reluctant to take the risk of appropriating the baits of young buffaloes which he

**"Waiting for Father"**. Drawn by Robert Armitage Sterndale. *Denizens of the Jungles*, 1886.

This illustrates the heart-rending story of a family of sloth bears in the Seoni hills. The father was the first to fall victim to the hunter. He is shown at the lower right corner of the picture being borne back to camp, while the mother and two cubs wait in vain for him to return. The next morning the mother and one cub met the same fate, while the second cub was captured by the hunter.

frequently came across; and, whereas formerly he expected no danger to lurk in the trees above him, now the need for circumspection was doubled by the possibility of a hunter being hidden in any leafy tree.

The tiger lay one night on the borders of a jungle clearing where the unfertile soil was covered with a growth of thorny bushes, which assumed strange shapes in the transparent gloom of the night. He had come for miles through the darker forest, moving slowly with the greatest circumspection; at each footstep the soft-padded paws seemed to feel the earth before any weight was allowed to bear on them; mechanically in their descent they pushed softly aside any dry leaf or twig which might, by their crackling, give notice of movement in the jungle, and now, tired from the constant nervous strain, he was resting before resuming his solitary way. A movement in the fantastic outlines of the bushes caught his attention, and he shrank still further into the friendly earth, all his fears at once aroused. A family of sloth-bears were feeding on the wild berries, embracing the bushes with shaggy arms, tearing off the fruit, intermingled with leaves and twigs, in the rough manner common to these beasts.

The tiger was glad even of this companionship, for it suffered him to relax his attention, for bears have the keenest power of scent, relying on this rather than on hearing or sight, so that timely warning would be given of any intruder.

The bears roamed round the little clearing, leaving no bush till despoiled of the fruit it bore, then commenced digging for roots and snuffing at the anthills to discover whether or not these were in occupation. The male bear soon found one to his liking, and commenced digging with his powerful claws to force an entry to the main passage, while the mother and her cubs sat around regarding the proceedings, though they could not possibly expect any share in the spoils. By dint of hard labour the bear had dug some three feet below the surface of the soil, and now inserting his muzzle in the tunnel drew deep inhalations which dragged with them crowds of unwilling insects into the moist mouth which was ready to receive them. The bear presented a ludicrous sight with his head buried in the earth and his hindquarters raised high towards the sky, and the noise of his breathing sounded loud through the still forest. After a time he commenced again to dig till he reached the nest, with its paper-like combs full of helpless maggots, and this he devoured in great mouthfuls. Then, while enjoying this selfish meal, he suddenly caught the scent of the tiger in the night air, and as quickly turned to fly. In his clumsy way he stumbled against the she-bear, and she, with the prompt retaliation of her tribe, at once struck and bit at her mate. Immediately the

**Tibetan Snowfinch >**
*Montifringilla adamsi*

Above, grey-brown with darker streaks on back. Wing dark brown with large white patches. In the tail, central feathers dark brown while outer are white, tipped with dark brown. Below, cream colour. Sexes alike. Usually very wild; sometimes amazingly tame and fearless. Keeps in pairs in breeding season, in small flocks thereafter, and in huge flocks of 2,000 or 3,000 birds in winter. Feeds on the ground, on the edges of melting snow patches, running like a lark – not hopping. Flight very undulating. During the breeding season, the male launches himself from a hill slope and with wings outstretched like the letter V, white tail feathers outstretched like a fan, hovers for a while, and then gently descends to earth, uttering a short plaintive song (Ludlow, *Ibis* 1928). Also displays on ground, loosely waving extended wings and jerking outspread tail. Fairly common resident, subject to vertical movements. Breeds in Ladakh, Spiti, NW Nepal, Northern Sikkim, and N. Himachal Pradesh, between 3,500 and 4,900 m. Affects high plateaux, boulder-strewn hillsides, and neighbourhood of upland villages (Sálim Ali and S. Dillon Ripley).

Tibetan Snowfinch *Montifringilla adamsi* Adams
*Birds of Asia*, Vol. IV, Parts XIX–XXIV, by John Gould, 1867–72. Painted by John Gould & Henry C. Richter.

Courtesy Mastek Foundation

Spectacled Finch (Red-browed Finch) *Callacanthis burtoni* (Gould)
*Birds of Asia*, Vol. I, Parts I–VI, by John Gould, 1850–54. Painted by John Gould & Henry C. Richter.

In Memory of Dosa Naoroji, from Rishad Naoroji

forest re-echoed with loud discordant cries, and the whole family disappeared into the forest, biting and scratching, in the belief that some enemy was amongst them seeking their lives.

The tiger wandered onwards through the jungle. He was now a different animal to what he had been in the days of his youth, when food was plentiful and danger not incessant. Now, for no fault of his own, he was proscribed; a price was set on his head, he was fired at on sight, and the very scarcity of food was used as a means to lure him to destruction. He was forced to satisfy his hunger by means he had formerly despised. He would lie by the drinking pools in the hot weather and ambush the jungle tribes while they were quenching an intolerable thirst; he would follow the females encumbered by the care of their young and profit by their maternal instincts to slay them; and would drive less powerful animals off their "kills" and appropriate the spoil. Domestic cattle he killed without mercy, so that he was known and dreaded throughout the countryside; he was always fierce and morose because he was at war with mankind, who had robbed him of his hunting grounds and with them of his means of living and of his contentment.

It was in these unhappy circumstances that his second courtship began, but on this occasion he forced a fight on his rival; for in the first place he was more savage than of old, and in the second it could not be tolerated that another should hunt in a forest where food was already but too scarce. Thus ill-temper and fear of dispossession urged him more than passion and in result there was a combat unique in its ferocity. There was no interruption from human beings, as these had mostly left the forest at the commencement of the malarial season, and the few foresters who remained were careful not to approach the spot whence the sound of the struggle proceeded.

The opponents were well matched, for what the stranger yielded in weight he gained in agility, and any deficiency in experience was outweighed by his impetuosity. The two rushed furiously at each other, meeting with a shock that seemed to compel them to stand upright, and in that position each tried to grip the other's throat and was repulsed by the powerful claws which scratched deep into the flesh. They retreated breathless, again and again to renew the attack after lengthening intervals; meanwhile the earth was trodden down and became slippery with moisture, though scored by the sharp claws of the hind-feet of the combatants. It was after many rounds had been fought, without marked advantage to either side, though both had received painful wounds, that the tiger slipped as he was repelling a specially violent onslaught by the stranger, and, over-borne, was hurled on to his back. In an instant the other rushed in to end the fray with teeth buried in the chest or throat of his foe; and here he made the mistake which cost him his life. He should have waited for the defenceless moment when the other was attempting to rise, instead of attacking him when in a position assumed by all the cat-tribe in moments of difficulty. And so it was that, before a grip could be secured, the stranger's head and neck were seized in a vice and at the same time his belly was ripped open by the hind-claws of his prostrate foe. His only wish was to be free of this deadly embrace, and at last he was allowed to stagger away mortally wounded. The crushing blow which followed seemed to drive the life out of him, and he had no feeling for the fangs which penetrated heart and lungs. The tiger lay long by his defeated rival; he was marked with scars which lasted to his dying day, he was sore with bruises and bites, and weak in everything but ill-temper and ferocity. It might have been better for him if he had ended his life at this time, if he had assumed no fresh domestic responsibilities, for the future was to bring even more bitterness than the past.

From *The Life of a Tiger* by S. Eardley-Wilmot, C.I.E., 1911.

**ARTICLE AND ILLUSTRATION COURTESY THE TATA POWER COMPANY LIMITED**

---

**< Spectacled Finch**
*Callacanthis burtoni*

Male: forehead, eyebrows, and round eye ("spectacles") crimson. Crown black; back brown; wings black spotted with white; tail black with white tip. Below, chin and throat pinkish red; sides of throat and cheeks black, rest fulvous brown tinged with pinkish red. Female has paler head with orange-yellow "spectacles", ochre-brown below, rest as in male. Keeps in pairs in breeding season, otherwise in small flocks. Feeds mostly on the ground, hopping about in the undergrowth, flying up to nearby bushes and trees when disturbed. Descends in ones and twos to resume feeding once disturbance has passed. Call: a loud, clear, high whistle *pewee* often followed by *pweu, pweuweu,* or *chipeweu*, all plaintive and melodious (Sálim Ali and S. Dillon Ripley), usually given from high up on a bare branch. Resident, Himalayas from Safed Koh and Chitral east through Hazara and Kashmir to Nepal and Sikkim, between 800 and 3,000 m.

# A Forest Fire

S. Eardley-Wilmot

At midday the forest was silent, only the insects and the birds were working and watchful, enlivened by the genial warmth. From time to time a great carpenter-bee passed with sonorous drone, or a wasp-like creature carrying a ball of moist clay to build a secure retreat for a posthumous offspring. A furtive ichneumon fly, encased in brilliant green armour, lay in wait for any chance to provide for its future family at the expense of another. On the pinnacles of the white-ant citadels the area of wet earth increased, shewing the activity of subterranean workers. Fly-catchers and bee-eaters perched silently on the boughs above. At intervals they flashed through the air and a flying insect was captured, or, escaping for the moment, was followed in its hasty flight by streaks of coloured light, azure or green, so rapid that the form of the hunter could scarce be distinguished. Then the bird returned to its perch with the insect in its bill, and paused to give thanks and to draw breath before proceeding with its meal.

At some distance a rustling became audible in the dry leaves; there was a sound of many pattering feet. The cold night breeze that blew down stream had long since ceased, now a warm breeze from the plains ascended the ravines towards the colder hills. A herd of spotted-deer appeared on the high ground and looked longingly at the clear water below, but before risking the descent, they paused to make sure that no hidden danger lurked. Most of the females had young fawns at their feet: they stood huddled together watching the movements of the old hind who was still intent on reading the signs of the jungle; and, though the sun shone brightly through the thin foliage above, yet from a short distance it was difficult to distinguish between the flecks of light, on the withered leaves and the white-spotted, chestnut hides of the deer. No sound or sight aroused the suspicions of the leader of the herd, but the breeze brought to her moist nostrils the unwelcome scent of danger, and she stamped fiercely with her fore-foot. Instantly the deer became motionless, and, after a moment's pause, turned away from the water and noisily disappeared into the forest. They reappeared higher up the stream and there, satisfied of safety, ran down to eagerly drink of its waters.

And now for the first time the stags appeared, bulkier and taller than the hinds, but meek of aspect, for on their heads were the tender velvet-covered growth of the new horns, and they feared the slightest accident that might lead to a painful injury. The stags had lost the fine colouring of the summer months, and were heavily coated in dull chestnut. The white spots were almost concealed by the long brown hair of winter. They moved with lowered heads to avoid the branches, and gave way before the more energetic females, not resenting even the gambols of the young fawns around them as they eagerly drank. From down stream came the warning cry of the red monkeys, perched high up in the trees, and at the first note the stags silently disappeared into the forest. The hinds, too, retreated from the water's edge and gained the higher ground; here they stood for some minutes considering what should be

**Mahratta Woodpecker >**
*Dendrocopos mahrattensis*

A small pied woodpecker. Above, brownish black irregularly spotted with white, also on wings and tail. Forehead and crown brownish yellow; small crest scarlet. Below, chin, throat, and foreneck white, rest of under-parts fulvous streaked brown, with a prominent scarlet patch on abdomen. Female similar to male but with the entire crown golden-brown without any scarlet. Both have long, stout, pointed bill and stiff wedge-shaped tail. Seen singly or in pairs, in groves and thin jungle. Affects open scrub country, light deciduous forest, mango orchards, and groves around villages. Scuttles up tree trunks in jerky spurts, tapping on the bark and digging into rotten wood for insects and grubs. The tail pressed against the trunk serves as a third leg to support the clinging bird. Distribution: throughout India up to 1,000 m in the Himalayas, also Sri Lanka, Pakistan, Bangladesh, Myanmar.

**Mahratta Woodpecker (Yellow-fronted Pied Woodpecker, Yellow-crowned Woodpecker)**
*Dendrocopos mahrattensis* (Latham)

*A Century of Birds from the Himalaya Mountains*, by John Gould, 1832. Painted by Elizabeth Gould.

Courtesy Toyota Kirloskar Motor Pvt. Ltd., Bangalore

Heart-spotted Woodpecker *Hemicircus canente* (Lesson)
*Birds of Asia*, Vol. V, Parts XXV–XXX, by John Gould, 1873–77.
Painted by John Gould & William Hart.

Courtesy HDFC Housing Development Finance Corporation Limited

done on this occasion of distant danger; then, led by the barren hind, they entered the forest and, in single file and noiselessly, took the way to some safer retreat till evening should fall.

Meanwhile the family of tigers, save when on the hunt, paid small attention to the movements of the jungle tribes around them; they lived their own quiet if laborious lives, and as the days passed the cubs throve apace. Soon they were able to play together on the soft sandbanks in front of the lair; they extended their knowledge by exploring the recesses of the tall grasses around the hollow log, and could even take refuge in this retreat on hearing a warning growl from the ever watchful tigress. But the presence of the family was long since known to all the jungle-folk. The crows were constantly scolding from safe perches in the vicinity, the jungle-fowl and the peafowl too, resented the intrusion, and the monkeys were never tired of proclaiming aloud what all around were already aware of; during the hours that should have been given to slumber after arduous hunting it was trying to be under the observation of all who might be idling in the tops of the overhanging trees.

Abroad, too, game was getting scarcer, and each week longer excursions were necessary to secure food, while the animals were so constantly on the alert that success was not so frequent as always to relieve the pangs of hunger. The strain on both parents showed itself in the growing ill-temper of both tiger and tigress. They still profited by each other's hunting, but the tigress showed the greatest caution in approaching any animal she had not herself killed; and on one occasion the tiger had so far forgotten the law of the jungle that he had driven her off the body of a pig she was ravenously devouring, and had appropriated it for his own repast. Everything, in short, indicated that a change to some other hiding place would be desirable where the presence of so many tigers would not be so notorious; and the more so that the grasses were commencing to dry up, that the trees, no longer afforded sufficient shelter against the sun, while the supply of running water began to diminish. More than once, too, the approach of spring was indicated by an ominous glow at a distance towards the south, which proclaimed the fact that the grasslands outside the forest were being burnt, and that at any time fires might extend to the forest itself, when it would go hard with any animal which could not outstrip the flames.

The uneasiness felt by the tigers was soon justified, for as the season became hotter and drier the evening dews failed to extinguish the fires which were lighted by cattle graziers during the day and it was not long before the nearer approach of the flames was heralded by clouds of smoke passing over the forest, and by flakes and spirals of burnt grasses whirling and scattering in the air. To the cubs, with their instinctive dread of fire, these were awful portents; they wandered to and fro in an aimless attempt to escape from their own fears, and at last sought refuge in the hollow log, where they huddled with piteous whining. Meanwhile the tiger had already slunk away down wind, heading for a swamp that afforded a safe protection from fire, save in the driest seasons; and the tigress, longing to follow his example, still hesitated to leave her young to their fate. She had indeed started on the road to safety, but finally returned to the lair and encouraged her cubs to follow her as she retreated from the rapidly approaching conflagration. But their strength was not equal to the occasion, and after a short time their tottering feet refused to bear them further over the obstacles they encountered in the forest; and finally, weak from fatigue and fear, they sat side by side, helplessly awaiting their fate.

The fire was now close at hand, near enough to light up the forest so that each stem stood darkly defined against its flow. From time to time the flames died away, and then reviving shot twenty or thirty feet skywards; or, flattened by the wind, covered the earth with a horizontal sheet of fire, licking up everything before it. As the heated blast reached the green foliage it wilted away and hung lifeless from the branches,

### < Heart-spotted Woodpecker
*Hemicircus canente*

A small squat black and buff woodpecker with a short thin neck, large crested head, and a stumpy rounded tail. In male forehead, crown, and crest black; back black with a broad buff band on either side marked with heart-shaped black spots. Below, chin, throat, foreneck, and sides of neck buffy white; rest of under-parts dusky olive and black. Female similar but with forehead and crown buffy white, crest black as in male. Affects moist deciduous biotope, partial to teak and bamboo forest. Frequently seen in pairs, jerkily creeping along the terminal twigs of tree, tapping on them to dislodge lurking insects. Call: a harsh but pleasant trill of 7 or 8 *twee, twee, twee* notes. A sharp double *tchlik-tchlik* in flight. Also drums on branches during the breeding season. Distribution: peninsular India from Kerala north to Gujarat and east to West Bengal, NE hill states, Eastern Ghats, and Bangladesh. Plains and foothills up to c. 1,300 m.

and the next instant a tongue of flame sprang forth and the tree was enveloped in a column of fire. When it passed only a blackened trunk was left to mark where so shortly before a monarch of the forest had stood in the pride of wide-flung leafy branches. The air seemed full of uproar and confused noises; the bamboos split with loud reports as of musketry; falling branches flashed and struck the earth with a crash of leaping sparks, and above all sounded the incessant roaring of the burning grasses, at one moment standing erect, and the next dissolved into clouds of bitter smoke.

From the direction of the fire came a stream of terrified animals running before the wind, each intent on its own escape, some so bewildered that they turned from imaginary dangers to rush blindly towards the flames; troops of monkeys climbed the tallest trees in the hope of reaching safety, and falling one by one, suffocated by the heated fumes, were dead before the fire passed over their helpless bodies. Above the tawny smoke which rose in billows to the sky king crows and other birds hovered, snapping up the insects which had escaped from the furnace below; they alone of the jungle folk showing no trepidation at the ruin that was overwhelming both animal and vegetable life.

Fear for herself and anxiety for her young seemed to rouse in the tigress a fury that was indifferent to all dangers. Missing her cubs she returned and endeavoured to carry the two smallest in her mouth and save them from the approaching flames. At times she succeeded in doing this for a few yards, then one would fall from her uneasy grip, and so she returned to and fro on the line of retreat until some distance was placed between them and danger. The third and strongest cub, unwilling to be left alone, also resumed his painful march, following instinctively the trail left by his mother; till after passing through a belt of shady undergrowth he stumbled down the bank of a stream and lay in its waters panting and exhausted. The crackling of the flames was still audible and the acrid smell of smoke filled the air, but the fire died

**"Deer of the Ganges"**. Drawn by George Trigger. *The Bengal Sporting Magazine*, Vol. XV, No. 85, March 1840.

away as it reached the evergreen shrubs lining the water-course, so that safety was at last assured.

Along the banks of the stream were many refugees, and the common danger made them indifferent to company otherwise distasteful. A truce was proclaimed that none might break, and the cub then saw more of the jungle tribes than before in all his short life. There were groups of spotted-deer who pressed against each other in their anxiety; the fawns had lost all the joyous abandon of infancy and the stags stood moodily with lowered heads, the grey velvet of the new growth of horn glistening in the yellow light. Sambhar hinds and calves also there were, silently watching, turning their huge ears to every sound; while one stag rested with heaving sides, a slow trickle of blood descending from his head, shewing that he had avoided the flames with difficulty and had lost his antlers in his headlong rush through heavy cover, perhaps some days before they were quite dead. Some pigs moved uneasily, snuffing in the low foliage, not desirous of the approach of even their own kind, and losing no opportunity to strike a petulant blow at any intruder. A barking deer stepped daintily over the soft ground; his little horns stood on pedestals above his head and these formed a V-shaped marking on his face; the two white tusks, that look so harmless and are so effective, gleamed from the sides of his mouth. Above were monkeys, brown and grey, that were for the time dull and silent after their exertions; not so exhausted, however, that they did not notice the tigress returning in her tracks, or fail to give warning to those below. The deer moved uneasily to one side as the tigress passed, and she, faithful to the law of the forest, turned neither to right nor left; she went straight to the shallow water where her cub was lying still too tired to move, nor did she show ill-temper when he scratched, and bit, and struggled against her interference. She seized him by the back and bore him gently away to the new home she had chosen for her family.

From *The Life of a Tiger* by S. Eardley-Wilmot, C.I.E., 1911.

ARTICLE AND ILLUSTRATION COURTESY BAJAJ AUTO LTD.

# Tigers of Salsette Island

JAMES FORBES

The summit of this wonderful mountain [near the Kanheri caves] commands an extensive view; the island of Salsette appears like a map around the spectator, presenting a fine champaign of rice fields, cocoa [coconut] groves, villages, and cattle; woody hills and fertile vales: the surrounding mountains form a fore-ground of grey rocks, covered with trees, or hollowed into gloomy caverns, the haunt of tigers, serpents, bats, and bees, in immense swarms; the horizon is bounded on the south by

**"Scene of a Melancholy Event on the Island of Salsette"**. From a drawing by Baron de Montalembert, 1774. From *Oriental Memoirs* by James Forbes, Vol. II, 1813.

**"Hindoo Peasant Ascending the Cocoa Nut Tree to Draw the Tari or Toddy"**. Drawn by James Forbes, Bombay 1768. *Oriental Memoirs*, Vol. I, 1812.

"The coconut palm *Cocos nucifera* is cultivated in damp hot regions of Asia, in low sandy situations near the sea. The tree yields toddy, a mildly intoxicating drink. The Toddy-drawer ascends the tall trunks without any apparent fear. On reaching the top he selects a spathe which is ready for tapping. He cuts the point transversely and fixes it in a curved position so that, when he has crushed the exposed flowers at the end the juice may flow freely. This is repeated daily…In a week or two the tree is ready to yield toddy…It is generally believed that a drink each morning is beneficial to the health." (*Flowering Trees and Shrubs in India*, by D.V. Cowen, 1957.)

the island of Bombay with the harbour and shipping, east by the continent, north by Bassein and the adjacent mountains, and west by the ocean. In various parts of Salsette are romantic views, embellished by the ruins of Portugueze churches, convents, and villas; once large and splendid.

The enjoyment of the picturesque and fertile scenery of Salsette is interrupted by the tigers which infest the mountains and descend to the plains: they not only prey upon the sheep and oxen near the villages, but sometimes carry off the human species. During our short stay, a poor woman gathering fuel on the skirts of a wood, laid her infant on the grass, when a tiger sprung from the cover and carried it to his den, in sight of the wretched mother!

Another of these ferocious animals prowling in a garden near Tannah [Thane], the capital of the island, suddenly put his head and fore-feet through the small window of a summer house where a friend of mine was sitting. Alarmed at his danger, he kept his eye stedfastly fixed on the enemy, rightly judging that the aperture was too small for the admission of his body; the gentleman then ran speedily to the house, and returning immediately with two or three armed servants, shot the monster through the heart, he having never moved from the spot.

From *Oriental Memoirs*, Vol. IV,
by James Forbes, 1813.

JAMES FORBES' account gives an idea of how freely tigers and other wild animals roamed the forests around Bombay during the later part of the 18th century (around 1780).

According to Hector MacNeil, there was abundant wild big game in the immediate neighbourhood of Bombay, towards the end of the 18th century: "The Governor and most of the gentlemen of Bombay used to go annually on a party of pleasure to Salsette to hunt wild boar and royal tiger, both of which were found there in abundance. In 1806 two tigers were seen near General Macpherson's bungalow at Kurla, while only a few days previously two persons were carried off from a village just a little further north, presumably by the same animals." Coming to the island of Bombay itself, we find the following record of 9th February 1822: "A tiger on Malabar Hill came down, quenched its thirst at Gowalia Tank and ran off over the hill between the Hermitage and the Prospect Lodge. Prints of its feet were distinctly visible this morning." Bishop Heber speaks in 1829 of "the occasional occurrence of hyaenas in Bombay," but the *Bombay Courier* of the same year records the sudden appearance of a tiget at Mazagaon: "The animal had apparently swum across the harbour and landed near the ruined Mazagaon fort, and thence was driven to the compound of Henshaw's bungalow where he was eventually shot by the guard of the Dockyard and certain Arabs." On 3rd March 1858, "Some officers of the P&O steamer *Aden* observed a tiger swimming from the mainland to Mazagaon. A boat was lowered and the crew shot the tiger." The same year a young Portuguese shot a tiger at Mahim. On 15th February 1859 a panther was seen near Kalbadevi Road and was afterwards shot by Mr. Forjett, the Commissioner of Police, near Sonapur – Professor P.R. Awati, *Bombay Past and Present*, souvenir of the 13th meeting of the Indian Science Congress, Mumbai, 1926, reprinted in *Sálim Ali's India*, 1996.

ARTICLE AND ILLUSTRATIONS COURTESY GOODLASS NEROLAC PAINTS LTD.

**Snow Pigeon** *Columba leuconota* (Vigors)
*A Century of Birds from the Himalaya Mountains*, by John Gould, 1832. Painted by Elizabeth Gould.

Courtesy Asian Star Co. Ltd.

# Goruckpore Terai, where Tigers were "Plentiful as Blackberries"
## Before the march of civilization made inroads on their domains

"Fleetwood"

### < Snow Pigeon
### *Columba leuconota*

The Snow Pigeon of high altitudes, well-known in both the Eastern and Western Himalayas, has a large proportion of the plumage white. In overhead flight the white body and blackish head are prominent. When flying below observer's level the blackish head, brown back, white patch on the lower back, grey wings with three dark bars, and blackish tail with white subterminal band are diagnostic features. Sexes alike. Affects rocky cliffs and gorges in the alpine zone and above snow-line. Lives and roosts in colonies on cliffs, small parties flying out to glean on grassy slopes or at edge of melting snows; in summer also in barley fields around Himalayan upland villages. Pilgrims visiting the holy cave of Amarnath in Kashmir always find them flying in and out. Resident in the Himalayas between 3,000 and 5,000 m; descending to low altitude (c. 1,500 m) in severe winter; does not extend into the desolate Tibetan Plateau.

In former times there was no district in India more justly celebrated for its tiger shooting than Goruckpore; year after year parties proceeded to its apparently inexhaustible Terai, and always returned with the spoils of numerous of these magnificent animals; in the time of Sir Roger Martin there was a well known tiger on the Nepaul frontier, who was the terror of the neighbouring villagers; man was his food, and he scorned to prey on any inferior animal. He once attempted to enter the hut of a Tharoo [tribal of the Indo-Nepal border region], but the inmate received him with such a blow on the head from a Jungle axe, that the Tiger was glad to retreat, and carried the scar of the wound to his dying day: by this scar he was known and recognized, and his depredations at last became so serious that Sir Roger Martin went out for the express purpose of killing him: he shot forty-eight tigers before he fell in with the one he was in search of, but the forty-ninth was "Le Balafre", himself, who fell, fighting to the last, and well supported his former character for ferocity.

Abbye Singh, the Rajah of Omorah, a very old and good sportsman, is known to have been at the death of nearly five hundred tigers, but it would be an endless task to enumerate all the instances in which large numbers of these animals have bit the dust and yielded their skins as trophies to the sportsman. From the above it can easily be imagined that Tigers were "plentiful as blackberries" a few years ago, and it is easily accounted for; after the Nepaul war, the Terai was one wide inhospitable waste, without a vestige of inhabitants or cultivation; intersected by nullahs in every direction, and abounding in swamps; the jungle sprung up luxuriantly and became the haunt of innumerable tigers and wild animals of all description. The annual inroads of sportsmen and some slaughter by Native Shikarees did not apparently much diminish the number of the tigers; year after year they were killed in the same spots, and it appeared that a desirable covert was no sooner vacated by the death of one, than another took possession of it, and till lately a party was tolerably sure of good sport in the Goruckpore Terai.

It is very different now; the tigers would never have been extirpated by sportsmen, – indeed those killed bore such a small proportion to the whole number, that for some years no decrease appeared to have taken place; but it is the change in the Terai itself which has diminished them, and which bids fair in a year or two to put a stop to tiger shooting entirely. Where formerly there was a howling wilderness, now villages have sprung up, the land is cultivated, and the jungle cleared: the march of civilization has made inroads on the domains lately sacred to the tiger and the bear; English Gentlemen have received grants of land from government on clearing leases, and by their efforts the jungle is rapidly disappearing: nor are the Nepalese behind us; on their side of the boundary, cultivation is increasing in a great degree, and the country daily becoming more densely populated. On the spot where we killed a fine tiger last year, there is now a thriving village:– the natural consequence of all this, is, that the tigers have been obliged to retreat before the approach of civilization, and where, formerly, there were hundreds, you now will not find one, they have taken refuge in the lower range of hills where they still find cover, but of course cannot be got at there: one may now pass through the Terai for miles without finding a good covert, and the only shooting, you are likely to see, is a Revenue Surveyor "shooting the sun" with his theodolite, and you, Mr Editor, instead of an account of the deeds of a Tiger party may expect to receive a report on settlements.

The poor Bunterrias are in despair. Never, they state, could they have believed that such an alteration could have taken place in the Terai in so short a time, unless they had seen it. Goshayun, one of the best of them may now exclaim

farewell
The spirit stirring roar, the ear-piercing trump,*
Pride, pomp and circumstance of glorious hunting
Farewell! Goshayun's occupation's gone.

Instead of putting up tigers we put up boundaries, and it is my opinion that next year there will scarcely be a tiger left in Goruckpore Terai. I sincerely hope that my prognostication may prove incorrect, but I very much fear the contrary.

**Red-headed Trogon >**
*Harpactes erythrocephalus*

In the male, head, neck, and breast deep crimson, sometimes with traces of a white breast-band. Back and upper parts rusty brown. Wing coverts and tertiaries finely vermiculated black and white. Long, broad, truncated square-ended tail black and white. Under-parts brighter and lighter crimson. Female differs only in having the head, neck, and breast dull orange-brown. Silent, sluggish, and rather crepuscular. Seen perching upright on snags and tree-stumps along shady jungle paths below the foliage canopy, occasionally making sorties after flying insects and seeking a new perch after each capture. Also eats berries and leaves. Usually silent. A brief mewing *cue* like an oriole repeated up to ten times. Resident in the Himalayas from Kumaon eastwards. Also NE states and Bangladesh.

**"What I Saw One Morning"**. *Records of Sport in Southern India* by General Douglas Hamilton, 1892.

**Red-headed Trogon** *Harpactes erythrocephalus* (Gould)
*Birds of Asia*, Vol. III, Parts XIII–XVIII, by John Gould, 1861–66. Painted by John Gould & Henry C. Richter.

Courtesy Suresh S. Kothari & Family, New York, USA

This must appear extraordinary after the good sport which parties had last season; but the change which one year has made in the appearance of the Terai is quite astonishing, and must be seen to be understood: a party of three gentlemen went out for 20 days last month, and "with all appliances and means to boot" only killed five Tigers; a second party is now in the Terai, but I have not heard what success they have met with.

It is a melancholy prospect to sportsmen in this part of the country, that of losing such a magnificent sport, though, if one reflects for an instant on the damage done by tigers, it is impossible to regret their decrease. There can be no question that the high and palmy days of Goruckpore tiger shooting may be considered as gone, never in all probability to return; occasionally a tiger may come down near the station and afford an hour's sport, and one or two may probably be picked up in the Terai next year, should any one think it worth his while to go there for them; but whoever expects to find sport which will compensate him for the risk and "*desagremens*" attendant on a tiger shooting expedition, may make up his mind to be most grievously disappointed.

Camp Hussunpore, 6th April 1837.

From *Bengal Sporting Magazine*,
Vol. IX, no. 41, 1st May 1837.

\* Of the elephant.

**Malabar Trogon >**
*Harpactes fasciatus*

A brilliantly coloured bird with a long, broad, and curiously truncated graduated tail. Male: head, neck, and breast blackish brown. Under-parts brilliant crimson-pink with a white gorget dividing them from the breast. Back and upper-parts yellowish brown. Wing blackish with fine wavy white barring. Female duller, with the under-parts orange-brown instead of crimson. Arboreal. They perch upright and have a knack of keeping their sober-coloured back turned to the observer. The broad, square-ended tail, as broad as the body, makes the motionless bird look like a strip of wood and difficult to spot until it flies (Sálim Ali). It prefers dense jungle but is also sometimes seen in the open. It is a silent bird and has, according to Davidson, both a cat-like mewing note and one which sounds like *kur-r-r* with all the r's rolled together. Distribution: peninsular India south of Gujarat and Khandesh, Western and Eastern Ghats and Sri Lanka up to c. 1,800 m. Ferguson says, "not uncommon in heavy forests of Travancore from 1,000 feet upwards."

ARTICLE AND ILLUSTRATION COURTESY G.A. RANDERIAN (P) LTD., KOLKATA

Malabar Trogon *Harpactes fasciatus* (Pennant)
*Birds of Asia*, Vol. I, Parts I–VI, by John Gould, 1850–54. Painted by John Gould & Henry C. Richter.

Courtesy Mahindra & Mahindra Ltd.

# Wounded Tigers, Panthers, and Bears

Reginald Gilbert

*Read before the Bombay Natural History Society on 10th July 1894*

The occurrence of so many accidents out shikarring this hot season has induced me to write a paper on the above subject, in the hope that it may be of use to those who may find themselves under the necessity of following up a wounded beast on foot. My experience of shooting in India extends over 17 years.

My remarks are intended only to apply to wounded tigers, panthers, and bears. On this side of India, it is almost impossible to get the use of an elephant, and nearly all big game shikarries are compelled to follow up wounded animals on foot.

On two occasions – once in Rewah and once in the Hyderabad State – I had the use of an elephant. I never would go out shikarring without one if I could possibly help it. I know nothing grander than following up a wounded tiger on a good staunch elephant. From a position of perfect safety you are able to behold all the grandeur of the charge of an infuriated tiger, and to have all the thrill of the sport without the danger of it, or, to quote the immortal Mr. Jorrocks, "all the spirit of war with only five per cent of its dangers." To those, therefore, who can obtain an elephant, I say never follow up a wounded beast without getting into the howdah.

The weapons of the tiger and the panther are not only their teeth. The sharp retractile claws, which they fix into the man attacked, render their victims almost powerless. The blow, too, which the tiger can give with his fore-paws will almost brain a man. A wounded panther, which I lately shot on the top of one of my men, made a large hole with his claws in the man's chest to the lungs. This man was bitten and clawed all over as well and died in a few hours, but the wound in the chest through to the lungs was the worst. This panther never left the man from the time he was seized till I shot it, and when I arrived on the scene (some 3 minutes or so after he was seized), the panther was worrying and shaking the man like a dog does a rat. The Indian bear, too (*Melursus ursinus*), can do terrible mischief with his claws. I have understood that they always try and tear the flesh off a man's face and shoulders with their claws; at any rate, in the only instance I have seen a man attacked by a bear, the bear tore one of the man's cheeks off besides clawing and biting him in other places so badly that the man died in a few hours. I have often had beaters knocked down or clawed in the beat by unwounded tigers and panthers when breaking back, but I have never seen a man seriously injured by an unwounded beast.

The question is, when the beast is wounded what should be done? Some may say leave him alone, but if this is done the next innocent native wood-cutter or herdsman who should be so unfortunate as to come near the beast will get killed. I need only mention as an example the case of my friend, the late Mr. G.L. Gibson, of the Bombay Forest Service. He wounded a tiger in Khandesh, but darkness coming on he had to leave it. Next morning he went out after it, and, close to the place where he left the tiger the previous night, he came across the body of a native boy who had been tending cattle and had just been killed by this wounded tiger. Whilst examining the boy, or shortly afterwards the tiger rushed out, seized Mr. Gibson, and from the wounds received he died. The plan I have always adopted in following up a wounded beast is to get as many beaters as I can and form them into a solid body. The whole body then move slowly forward. It is necessary to move slowly and carefully, and occasionally to turn round to see that the men are not straggling. Every now and then I make the beaters throw stones forward into the thickest jungle and I encourage them to make a noise. The wounded tiger or panther more often than otherwise reveals his presence by growling. I have known, however, a wounded tiger which allowed me to get within eight yards of him in Karve reeds without growling. I discovered his position there by hearing his heavy breathing caused by a lung wound and by the movement of the reeds. My theory is, that a wounded beast will not charge home into a solid body of men. He may commence his charge, and if any of the men were to rush out in terror from the main body, it is probable the animal might seize him, but I think if all stand firm he will not come on. Native beaters seem to be great believers in this theory. I have often known them, when a tiger breaks back or roars near them, gather together in a mass and shout for the purpose of keeping the beast off them. It may be said that one ought not to allow the beaters to run any risk as they are unarmed. It would not be right to force beaters against their will to join in a following up, but in my experience I have always found a large portion of the native beaters very plucky, and often recklessly brave, requiring restraint and frequent words of warning. Here I am tempted to say in praise of native shikarries and jungle men. In this class I do not include the Bombay professional native shikarry. How often have I seen them unarmed do the bravest of deeds, which it would be brave even for a man armed with the best modern weapons to undertake. In following up especially, when I have had few men with me I have also adopted the plan of moving from tree to tree and then sending up a man to the top of the tree to spy around. I once wounded a tiger, which got in very high reeds; when we got within 40 yards of him he commenced to growl. I moved near a tree and sent a man up who reported he could see him. I climbed to the very topmost bough and from that position I could just see him and was able to kill him.

**"A Long Shot Just in Time".** *The Old Forest Ranger* by Captn. Walter Campbell, 1842.

In hilly ground I always work from above, so as to have the advantage of being on higher ground when the charge comes. There can be no question I think that a wounded tiger, panther or bear will charge home in the face of both barrels discharged point blank into him. He will make good his charge against one or two men, but not, as I said before, against a mass. Often, too, a wounded beast will commence a charge, but draw back when he sees his adversary facing him like a rock showing no fear. He will then swerve off. The soul-stirring growl he makes whilst charging is, no doubt, made for the purpose of causing fear, and in many instances the bravest of men quail when suddenly hearing it. In the late unhappy accident to Colonel Hutchinson, the Revd. E. Jenkins Bowen informs me the tiger charged from a distance of 30 yards, four guns in a line, but separated by an interval of two or three yards. Seven barrels were discharged into this tiger whilst he was charging – one by Colonel Hutchinson as the tiger was about to strike Mr. Bowen with his claws, and another barrel by Mr. Bowen, at a distance of three feet from the muzzle of his rifle, when the tiger was actually seizing or had seized Colonel Hutchinson; but in spite of this the tiger not only shook and mauled his victim, but carried him off some ten yards before he was killed. In this case the beaters were not with the guns: Mr. Bowen informs me he sent back all the beaters, thinking the occasion to be one in which no beaters were required. This, however, must have been an exceptionally fiendish tiger.

Now a word as to bears. These are the most foolish of animals. Wound a bear and he commences to fight with his companion if he has one. His brain-power is very deficient. After he has been marked down, walk up to his lair and wake him up. He takes five minutes before he even knows he is awake. When wounded, he will generally show fight, and I think he should be attacked in the same way as the tiger. After firing at him, if he still continues his charge, I recommend the sportsman to throw his sun-topee at him and then bolt. By an accident I found out a wounded bear would stop and claw up a topee instead of pursuing his enemy. I once, in thick jungle on the ghauts, fought a wounded bear nearly all day. On the first occasion I got quite close to him and in a second he charged out at me. I fired both barrels point blank into him and turned and fled. After going but a short distance – the bear following me – a branch knocked off my pith topee. To my surprise the bear stopped, seized my topee and smashed it to smithereens and chewed one end of it in his rage.

**ADDITIONAL NOTE**

Since the above paper was read, a friend has drawn my attention to Rice's book on tiger-shooting in India. The book, which is now very difficult to get, was published in 1857, and I see the author holds exactly the same opinions as I do. At page 57, after alluding to forming up the party, he says: "The whole party in a compact body keep close together, move at a snail's pace, yell with their utmost power, and create what really is a most infernal din. No tiger will face such a mass of men and noise as this. They sometimes charge down within even a few yards, but then invariably turn off. With this system there is perfect safety to every one, no matter how dense the jungle may be." It may be noted the author gives his party's bag in Rajputana for one year at 68 tigers killed and 30 wounded, 3 panthers killed and 4 wounded. Bears killed 25, wounded 26. Oh, ye gods!

From *JBNHS*, Vol. IX, No. 1, 1894.

**Purple Sunbird >**
*Nectarinia asiatica*

Garden bird abundant in the plains. A small bird with a long curved beak; male metallic blue-green and purple often looking black in poor light, female brown and yellow. Male in non-breeding plumage resembles female – brown to olive-brown above, pale dull yellow below – but with darker wings and a broad black stripe running down middle of breast. Pairs in lightly wooded country. Largely dependent on flowers for its food, it is attracted by the blossoms of flowering shrubs or trees, not only feeding on their nectar but also on the insects they attract. Usually perches on the stems of the plant, flits from flower to flower indulging in gymnastics to reach the desired food. Utters a sharp monosyllabic *wich, wich* while flitting about. The breeding male sings from exposed perches, *cheewit-cheewit-cheewit*. Distribution: throughout India, Bangladesh, Pakistan, Sri Lanka, and Myanmar.

ARTICLE AND ILLUSTRATION COURTESY SATISH D. CHOKSI, INDIAMCO

Purple Sunbird *Nectarinia asiatica* (Latham)
*Birds of Asia*, Vol. II, Parts VII–XII, by John Gould, 1855–60. Painted by John Gould & Henry C. Richter.

Courtesy Kanwal K. Grover, Hindustan Export & Import Corporation Pvt. Ltd.

**Fire-tailed Sunbird** *Aethopyga ignicauda* (Hodgson)
*Birds of Asia*, Vol. I, Parts I–VI, by John Gould, 1850–54. Painted by John Gould & Henry C. Richter.

In Memory of Dhirajlal N. Shroff, from Manjula D. Shroff, William & Rudra Connal

# The Indian Wild Boar as Grazier

A.K. NELSON

The following story illustrating the sagacity of the Indian boar is told by Mr. Lowrie: – "we were encamped at the village of Raitum close to some low-lying hills in a very wild part of the Raipur district, and had just finished tea under the shade of a *kusam* tree, and were indulging in a smoke, when I saw a fine boar run past the tents towards the fields; some of the servants were standing close to the tent, and I happened to ask one of them if he had seen the boar go by; yes, he said and what's more, that is the grazier to a villager living in the huts at the back of our camp. This we both exclaimed was too good a story, but I had hardly spoken when up came the boar, driving eight goats in front of him; we followed him right on to the village, and in went the goats into a hut, as it was shutting-up time. These eight goats were only part of the herd that had strayed out into the fields, the main lot having been penned before and the grazier missing those he now brought had gone and driven them in. He himself entered the hut last of all. On talking to the owner after this extraordinary occurrence, we were told that he had got the boar when quite a youngster, three years ago, and brought him up on goat's milk; as the young boar grew he was taken out by the villager's son along with the herd to graze. Last year in quite a matter-of-fact way the villager informed me that he had put the whole herd, consisting of five and twenty goats, in charge of the boar, and right well had he managed his charge. They are all let out at six in the morning and away goes the grazier with his charge into the jungle,

**< Fire-tailed Sunbird**
*Aethopyga ignicauda*

Small bird, like a sparrow in size. Male, adult: crown metallic purple above. Sides of crown, nape, back, and tail all scarlet. Rump yellow, wings olive, throat metallic purple, rest of the under-parts yellow, breast with orange tinge. The central pair of elongated tail feathers is bright. Female: olive and more yellow in rump and belly. Food: chiefly nectar but also insects and spiders. Song: a high-pitched *dzidzi-dzidzidzidzi* continuously repeated as the birds fly from bush to bush, pursuing each other. Common resident in the Himalayas from Garhwal east through Arunachal Pradesh; thence south through Nagaland, Manipur, Assam (Cachar hills).

**"Hors de Combat"**. Drawn by Robert Armitage Sterndale. *Denizens of the Jungles*, 1886.

Little Forktail *Enicurus scouleri* Vigors
*Birds of Asia*, Vol. III, Parts XIII–XVIII, by John Gould, 1861–66. Painted by John Gould & Henry C. Richter.

Courtesy Rahulkumar Bajaj Charitable Trust

and punctually at dusk they are brought home. We then enquired if any goats had been lost; this amused the old man immensely, as he told us that no one dare go near the goats while out grazing; it was as much as his life was worth, and what was more the old boar would not stand any stray animals from another herd joining his charge; these were soon snouted out. During the year the owner had never lost one of his animals by a panther, though there were a fair number of these animals about. No doubt the boar had looked well to their welfare; a goat has only to "ba" and he is there to see what the matter is. This marvellous boar, I am sure, could well hold his own at any competition of sheep-dogs in penning his goats. On leaving camp the next morning Col. Henderson and I went down to see the grazier drive off his goats into the jungles, and I then managed to get a couple of photographs of the boar and his goats.

From *Central Provinces District Gazetteers*
by A.E. Nelson, *Raipur District*, 1909.

### < Little Forktail
*Enicurus scouleri*

Above: forehead white; head, neck and upper back black. A white triangular bar across wing. Lower back white, interrupted by black band on rump. Tail short, blackish, slightly forked. Below: throat black, rest white. Sexes alike. Found at Himalayan torrents and waterfalls, running along the water's edge or hopping from stone to stone in search of food. Incessantly wags stumpy tail and rapidly opens and shuts it in rhythmic scissor-like motion. Resident, throughout Himalayan region between c. 800 and 3,300 m, from the extreme NW Frontier to Arunachal Pradesh and NE hill states.

ARTICLE AND ILLUSTRATION COURTESY MRS. PUSHPA KIRTILAL BHANSALI
ON THE OCCASION OF HER 90TH BIRTHDAY

# Khedda
## Elephant Trapping

"Master Edward"

The part of the country where the best elephants are to be got or where the best *cadies* are, is Komilla or Komillreeah across a range of hills three hundred miles to the N.E. of Chittagong. A party of five hundred coolies, ten shikaries or elephant catchers, ten jouse walahs or guides acquainted with the jungle passes, and twenty or thirty men armed with firelocks, together with a European superintendent, proceeded to this spot merely taking their own provisions; other necessaries such as tents, &c. being quite inadmissible.

On approaching the haunts of the elephants, great caution was required, for the least noise warns these sagacious animals of their danger, and is the signal for a general flight, knocking down large trees, and their track may be heard for a mile by the crashing of bamboo jungle with the report of a six pounder. Patiently the party traced them for seven or eight days until a favourable spot offered for ensnaring them. The main body halted, while a few of the jouse walahs stripping, proceeded in advance to watch the movements of the herd and shortly returned with a report of their number; for this knowledge is very requisite when two require as much trouble to snare as twenty.

The number of the herd being considerable, preparations commenced. The five hundred coolies under the direction of the shikaries and jouse walahs were divided into two parties, each taking a different route, and two men stationed at intervals until a circle is formed surrounding the herd. A whistle formed of bamboo answers the purpose of communication and is less likely to attract attention than the human voice.

The elephants are now hemmed in and a path through which they are accustomed to pass is sure to lead through the enclosure. At this path a double force is placed and with every two men a third with a firelock. Thus matters remain for a day or two; the whole body always remaining concealed, magnifying their numbers. Should the elephants endeavour to break through, shouts, yells and the sound of fire-arms deters them, and in the mean time a strong stockade is built,– the entrance opening on the path. The walls of the stockade, formed of magnificent teak or *jarool* trees, stand seventeen feet high, three deep, strengthened by cross-pieces fastened with rattan. Round the interior, a trench six feet deep, ten broad at the top, narrowing to four at the bottom, is dug to prevent the elephants, when caught, injuring the stockade, and the path leading to the entrance is enclosed for some distance by one row of the same, five feet high; the interstices stuffed with rattans.

Matters being thus arranged, a party is stationed round the stockade, and on both sides of the path, armed with spears thrust through the crevices to prevent the elephants rushing against the walls. The remainder then get in rear of the elephants, shouting, firing guns and making other hideous noises. Naturally alarmed, the elephants rush down the accustomed path or tract into the stockade; dried bamboos, leaves, &c., already prepared in the trench across the entrance is fired, preventing all egress, and the wall of the stockade speedily thrown up. Information is now sent into Chittagong and elephants and material for making rope required; for every thing is prepared on the spot: nothing, as I said before, starting with the expedition in the first instance,– the sagacity and scent of the elephant being proverbial. The shock caused by the rush of fifty wild elephants may be imagined, when large trees are rooted up, and the ground trembles as if affected by an earthquake. The ropes being prepared, four elephants (females, because males fight) are sent into the stockade, one carrying the ropes with three shikaries on her: the others have but one mahout each, and the whole dressed in black, crouch down on the neck to avoid observation. The herd frightened crowd together: one is selected and surrounded so closely by the tame elephants as to have little use of his limbs. A shikarie slips down, secures a rope round each hind leg, to which is attached another firmly fastened to a strong tree. In this manner, all in turn are secured, and left for two days until somewhat exhausted by vain efforts to escape.

Occasionally there is some difficulty, but the spears of the mahouts keep the refractory in check, while the sagacity of the tame elephants restrains their power. Should however the prisoner prove too restive, the shikarie slips away under the belly of the tame ones, mounts with a rope ladder and waits for a more favourable opportunity. Notwithstanding every precaution, accidents do happen, and in this case two men lost their lives. It must not be supposed our friends in the stockade are fasting, for should there not be water inside, it must be supplied and food also. The tame elephants now go in again, supplied with rope as before, and contrive with their trunks to cast a noose over the neck, of the prisoner, another over the body just under the

**Trapped Elephant**. *Sketches of the Natural History of Ceylon*, with engravings from original drawings by Sir J. Emerson Tennent, 1861.

shoulder, and a third over the loins. These two last may more properly be called braces; a guy from the neck forms a breeching connecting the braces to prevent the one round the neck becoming too tight, and the prize is led out secured to his captor.

In case of a large powerful male elephant being troublesome, two tame ones are required to secure him on each side, and sometimes three; the last brings up the rear, to which is attached a strong rope fastened to the hind leg of the wild one to check him if he moves forward, and should he stop, to assist the other two by a charge in the rear.

One would suppose that all trouble is over but their excessive timidity requires great care to get them into Chittagong safely, for their dread of wild beasts renders it necessary to prevent even a dog crossing their path; and a party, therefore is always sent on to warn the villages and to shoot all stray quadrupeds,– a precaution saving much expense and annoyance where, if one escapes, a month may be spent in retaking it.

The females are very careful of their young: in crossing nuddees and streams they see them safely over, and should it prove too deep, lift them with their trunks on their heads and swim over. At the birth of the young, the herd, male and female, station it in the centre and never leave it unprotected for a moment until able to shift for itself.

Many female elephants may be observed with short tails. It is thus accounted for, and certainly appears to me a good reason, though not generally allowed:

The male elephant when "*must*," in its mad fit seizes the female, and if opposed in its desires, revengefully bites off the tail. Thus, when the herd are in the stockade, many may be observed carrying their tails like a frightened cur to protect themselves from similar curtailments;– a pun but it must remain.

The Komilla elephant is very superior and is distinguished by its large carcase: powerful, well formed short legs, small broad head with the trunk well set up; broad at the root and long. The long bushy tail also appertains to the Komilla elephant and their height varies from nine to ten feet.

The Sylhet elephant has an ugly head, narrow short trunk, small and deformed body, falling off to the rear, and small legs.

This expedition was particularly successful, having caught in two months, fifty elephants for the service of the Honorable East India Company. The privations, however, must have been severe to cause the death of twenty followers accustomed to the work. Once in the midst of the dense jungle where neither sun nor air could penetrate, inhaling the malaria and living upon coarse food. No situation could be more dreary, and although the sport must be delightfully exciting, my narrator or rather informant, assured me that he had only known one gentleman to attempt it, and he thought once was enough.

From *Bengal Sporting Magazine*, Vol. VI, no. 23, 1st May 1836.

**Khedda, Trapping Elephants**
*Oriental Field Sports*, Vol. I, by Thomas Williamson, 1808. Drawn by Thomas Williamson & Samuel Howitt.

Courtesy Tata Sons Limited

# Lion-Hunt in Hurriana

"Quondam"

It is about twenty years since my regiment was stationed at Hansi, in the centre of Hurriana, which place is well known as having been the principal fortress of that extraordinary adventurer George Thomas.

The fort has since been much strengthened by our engineers, the town has considerably increased, and the canal has improved the appearance of the surrounding country by inducing trees to grow in its neighbourhood. The plain of Hurriana is like a vast sea; the hill of Tosham and, till lately, a peepul tree near Hansi, like ships in the offing, were the most conspicuous objects and could be seen from a great distance towering over the stunted bushes and grass jungle. The neighbouring district of Hissar is famous for its pasturage, and extensive plains, abounding with nutritious natural grasses. Hansi was then in its "high and palmy state," and considered the best sporting country in India.– Lions were found in considerable numbers, although lately

**"Lion and Lioness in Their Native Haunts"**. Drawn by William Kuhnert. *Harmsworth Natural History: A Complete Survey of the Animal Kingdom*, Vol. I, 1910.

they have become exceedingly rare; and a sportsman might have filled his bag with black partridge in front of the parade [ground]. My sojourn there formed the happiest period of my life.

A good number of black partridge are still to be found in the preserved grass of the stud department; the district is still famous for the stoutness of its hares, and I should think the banks of the Cuggur and the Batty country must still afford a good sporting tract, and where occasionally a lion may be met with. There are abundance of wild hogs, and the country is particularly safe for riding.

The first lion hunt I ever was present at was the most beautiful sight I have witnessed. The party assembled at Hissar, where some of the sporting elephants of the Marquis of Hastings' retinue were stationed. A Duffadar's party of Skinner's horse accompanied us. The presence of suwars in lion hunting is very necessary; the plains being extensive, the animal is liable to be lost after the first onset unless suwars are at hand to go out on the flanks or to push on ahead to mark the jungle the lion retires into. In general when a lion is pursued, he will either endeavour to get away by sneaking off or take to the open country and there await the attack; the latter a tiger is never known to do, and I consider it to form the only peculiar difference of the two kinds of sports. A lion that takes to this open fighting gives by far the most exciting sport of anything I have seen in tiger hunting and is the most trying for the elephants. Our party started from Hissar after an early breakfast, and although we had no particular information, we soon came to a jungle in which it was pretty certain the animal we were seeking was tenanted, as the whole population of a neighbouring village, large and small, of both sexes, had come out to see the sport, and anxious to have a good view posted themselves on an open and high spot near the jungle we were about to beat. Soon after entering the jungle the lion was put up and fired at;

"**A Hurriana Lion**". Drawn from a specimen living in the Tower Menagerie, London, 1820s. "The Lions of Asia" by R.I. Pocock, *JBNHS*, Vol. XXXIV, No. 3, 1930.

"**Fattypore Sicri**". Drawn by William Purser, sketched by Captain Robert Elliot. *Views in India, China, and the Shores of the Red Sea*, Vol. I, 1835.

the suwars, perfectly understanding their part, charged out from both flanks to watch him. To our astonishment the lion made directly for the villagers; it was impossible for us to give them the least assistance till our tardy elephants came up when it would have been too late, but two of the servants behaved nobly, and saved the crowd from the anticipated visit, for the villagers had already taken to flight and were hard pressed, when the first suwar rode close up to the lion, whose attention was immediately attracted and turned round upon him taking fearful springs, and was just about making a finale of the horse and rider, when the second suwar dashed in directly between the two: the lion now pursued him, when he reined his horse up, and waiting for the lion, cut him in the mouth with his sword, while, at the same instant, his horse bounded off at full speed and saved himself from the return of the compliment. The lion disappointed and foiled in his purpose, retired back to the jungle, where we followed and killed him at the first volley. He was a young but nearly full grown male, stood exactly three feet high and was nine feet long; his mane was nine inches in length; the cut made by the suwar's *tulwar* was about three inches long and two inches deep on his upper lip. The women of the village were exceedingly anxious to burn the lion's whiskers, which they did in spite of every remonstrance: in this part of the country it is done with a view to ensure connubial happiness, and they also hold it sound doctrine that offerings made to a male lion propitiates *muhadeo* in favour of barren women. To the eastward, tigers' whiskers are carefully burned because they are considered very poisonous; if a person could contrive to bolt pieces of them cut into lengths of a quarter of an inch they might irritate the stomach to that degree so as to cause death, otherwise, there can be no reason to suppose they are poisonous. The flesh of lions and tigers is esteemed by natives a good medicine in certain diseases; for this purpose it is dried and made into a powder, and the fat is very valuable for external applications.

From *Bengal Sporting Magazine*, Vol. I, 1833.

**Peregrine Falcon >**
*Falco peregrinus babylonicus*
A compact, pointed-winged, sleek falcon. Recognized by slaty-black crown, cheeks and moustachial stripes, contrasting sharply with white chin and throat. Slaty-grey upper-parts barred with black; under-parts white tinged with buff, spotted on chest, and blackish bars below; legs yellow; beak bluish. Sexes alike. Female larger. In overhead flight whitish compact body, black barred under-wings, unexpanded tail and deliberate flight are distinctive. Seen singly, usually hunting at dawn or dusk, swooping on prey with terrific force. Feeds on waterfowl, pigeons, partridges, and other birds. Winter visitor throughout the subcontinent.

ARTICLE AND ILLUSTRATIONS COURTESY THE INDIAN HOTELS COMPANY LIMITED

Peregrine Falcon (Red-capped Falcon) *Falco peregrinus babylonicus* Sclater
*Birds of Asia*, Vol. IV, Parts XIX–XXIV, by John Gould, 1867–72. Painted by John Gould & Henry C. Richter.

Courtesy Pheroza & Jamshyd Godrej

# The Past and Present Distribution of the Lion in India

N.B. Kinnear

There is no evidence to show that the lion inhabited Afghanistan or Baluchistan within historic times, but it was formerly found in Sind, Bahawalpur and the Punjab, becoming extinct round Hariana, in the latter province, in 1842. It was however extinct in Sind before that date and the last on record was shot near Kot Deji in 1810. Exactly how far eastwards the lion was a regular inhabitant we do not know, though there is a statement of one being killed in the Palamaw district, Behar and Orissa, in 1814, but whether this was merely a straggler or not, there is no evidence to show. The southernmost limit appears to have been the Narbada. In 1832 one was killed at Baroda, while further north it was comparatively common round Ahmedabad in 1836. Central India in these early days was one of the strongholds of the lion and to give an idea of its numbers we may mention that Lydekker was informed that during the Mutiny, Colonel George Acland Smith killed upwards of 300 Indian lions and out of this number 50 were accounted for in the Delhi district!

The occurrence of the lion in Cutch is doubtfully recorded. The lion probably was found in Cutch at one time but the records are not satisfactory. Lt. Dodd mentions that Burns about 1830 wrote that lions as well as tigers, bears and wolves were found north of Bhooj, but that none except the last named were now found, though a solitary lion was shot near Bela on the Runn, which was supposed to have been a straggler from Guzerat.

Edward Blyth, the curator of the Royal Asiatic Society of Bengal, in his catalogue of the mammals in the collection, which was published in 1863, wrote that the "lion was extirpated in Hurriana about 1842, a female was killed at Rhyl in Damoh district Saugor and Nerbudda territories, so late as the cold season of 1847–48, and about the same time a few still remain in the valley of the Sind river in Kotah, C.I. The species would appear to be now extinct in that district."

A few years later writing in the *Oriental Sporting Magazine,* Blyth drew attention to some more recent records of the lion, which he said must have come as a surprise to sportsmen and naturalists, as it was thought that they had been long exterminated in these localities.

These two records consisted of one from Deesa, where Lt. Clarke of the Royal Artillery was badly mauled by a lioness in March 1864 and lost his arm, and near Gwalior, where three officers, out shooting in March of the following year came

**Grey Treepie >**
*Dendrocitta formosae*

A greyish looking treepie when seen flying or in the forest, with a brown throat and small patch of chestnut covering the root of the tail. Forehead black, crown of head and upper back ashy; the rest of the back brownish-buff with the rump paler. The sides of the head, chin, and throat sooty-brown fading on the rest of the lower plumage. Wings black with a flashing white patch. Sexes alike. Seen in pairs or small parties keeping to heavy or thin forest areas. They are noisy birds and have the same undulating flight of other magpies. Usually they keep to trees but sometimes descend to the ground. According to C.M. Inglis their food consists of fruit, insects, birds' eggs, and probably young birds also. Jerdon saw one eating grain as well. Resident from Himalayas in N. Pakistan (Margalla and Muree hills), Himachal Pradesh, western Nepal to Bhutan, Arunachal Pradesh and south to Manipur, Mizoram, and Meghalaya, Eastern Ghats and Bangladesh. Occurs in tropical and subtropical moist deciduous and semi-evergreen forest between 600 and 2,100 m, and also in the well wooded plains of Assam and Northern Bengal.

Grey Treepie (Himalayan Treepie) *Dendrocitta formosae* Jerdon
From *The Journal of the Bengal Natural History Society*. Painted by C.M. Inglis.

To the Memory of Avatara Krishna & Shama Wattal

Rufous Treepie (Indian Treepie) *Dendrocitta vagabunda* (Latham)
*A Century of Birds from the Himalaya Mountains*, by John Gould, 1832. Painted by Elizabeth Gould.

In Memory of Jayantilal Chandulal Kothari of Palanpur, from Vinaben J. Kothari, Rekha & Ashok J. Kothari & Family

**"Maneless Lion of Gujarat"**. *Routledge's Picture Natural History* by the Rev. J.G. Wood with 700 illustrations by Wolf, Zwecker, Weir, Coleman &c. engraved by the Dalziel brothers, 1885.

**< Rufous Treepie**
***Dendrocitta vagabunda***
A long-tailed chestnut-brown bird with sooty head and neck. Black-tipped long tail and greyish white wing-patches conspicuous in flight. Flight undulating – a swift noisy flapping followed by a short glide on outspread wings and tail. Sexes alike. A strictly arboreal bird of open forest, often near gardens, usually in pairs, with a loud and melodious call *ko-ki-la* often mixed with a harsh rattling cry. Food: fruits, berries, insects, small snakes, caterpillars, lizards, also small birds and their eggs. Treepies are regular members of mixed hunting parties in the forest. Found in the whole of India from lower Himalayas to Kerala, and in Pakistan, Bangladesh, Myanmar, but not Sri Lanka.

suddenly on three lions, two of which they secured. Blyth seems to have missed certain records, for in 1863 Col. Martin of the Central Indian Horse, and Mr. Beadon, the Deputy Commissioner, saw and killed no less than eight lions at Patulghar, 70 miles north-west of Goona while in 1864 Mr. Arratoon of the police "shot at and wounded a lion near Sheorajpur (25 miles west of Allahabad) and eventually with native help stoned him to death as he had no spare ammunition." In 1866 Blanford tells us that Messrs. Lovell and Kelsay of the railway staff at Jubbulpore, shot a lion in Rewah near the 80th milestone on the railway from Allahabad to Jubbulpore, and in the same year no less than nine lions were shot by one party in the neighbourhood of Kotah, Rajputana.

Round Goona lions were still numerous and two or three were shot in 1867, and Blanford, writing in the Journal of the Asiatic Society of Bengal for that year, says "a few appear to be killed about Gwalior and Goona, but the animal is scarce." At the end of his article he summarized the distribution of the lion in India at that date as follows:– "The lion seems still to exist in three isolated parts of central and western India, omitting its occasional occurrence in Bundelkund. These are (1) from near Gwalior to Kotah, (2) around Deesa and Mt. Abu and thence southwards nearly to Ahmedabad and (3) in part of Kathiawar, in the jungles known as the Ghur."

On Waterloo day, 1872, Sir Montagu Gerard killed a lion on Cheen Hill, nine miles from Goona, and the last one in Central India proper appears to have been that mentioned by Sclater as having been killed by Col. Hall near Goona in the following year.

In Rajputana they became extinct about the same date and in the Gazetteer of the "Western Rajputana States Residency and Jodhpur Residency" we find that a full grown female lion was killed on the Anandra side of Abu by a Bhil shikari in 1872, and in Jodhpur "the last four" are stated "to have been shot near Jaswantpura about 1872."

Lydekker gives 1888 as the date the last lion was killed in Guzerat exclusive of Kathiawar, but the last record I have been able to find is that mentioned by Colonel Nurse in the Society's Journal, Volume XIII, 1900, in which he says "the last, I believe, killed in 1878 near the village of Bhoyen, about two miles from Deesa." According to the Gazetteer for Palanpur the lion was "now very rare" there in 1880.

The lion is still found in small numbers in the Native State of Junaghad in Kathiawar, where they are principally found in the Gir forest, but occasionally lions stray over the border into neighbouring states, where it is not long before they are shot.

For information in regard to the present position of the lion in Junaghad reference can be made to Colonel Fenton's two papers in the Society's Journal, Volumes XIX and XX, and Mr. Crump's notes in the Mammal Survey Report for Kathiawar in Volume XXII.

From the article "The Past and Present Distribution of the Lion in Asia" which appeared in *JBNHS* Vol. XXVII, No.1, 1st July 1920.

**N.B. KINNEAR** (1882–1957) was the son of an Edinburgh architect. He started his natural history career as a voluntary worker in the Royal Scottish Museum at Edinburgh under the able guidance of Dr Eagle Clarke, the Director, and a distinguished ornithologist. He came out to India in 1907 as the Bombay Natural History Society's first stipendiary Curator to organize and look after the considerable zoological collections, particularly of vertebrates, that had been amassed by its enthusiastic members from all over the erstwhile "Indian Empire" during the early days of the Society. Kinnear gave invaluable service to the Society by placing the whole of its museum on a sound scientific basis. He also provided more effective assistance to members of the Society who sought his help and generally guided the work in directions which produced greater scientific gains. Kinnear contributed greatly in the systematic survey of the Mammals of India, Burma and Ceylon which the Society had started. Kinnear who during his tenure as Curator also served as one of the editors of the Journal resigned his post in 1919 to take up a special appointment as assistant in the Bird Department of the British Museum. There he steadily rose to become Assistant Keeper of Zoology and then Keeper, finally ending up as Director of the British Museum in 1947. Kinnear was knighted in 1950.

**Changeable Hawk Eagle >**
*Spizaetus cirrhatus*

A slender forest eagle, dark brown above, white below with narrow black longitudinal streaks on throat and dark chocolate streaks on breast. The prominent black crest is diagnostic. In overhead flight, the longish tail, short rounded wings, white spotted body help in identification. Dr. Sálim Ali writes in *Birds of Kerala*: "It sits bolt upright on a branch near the top of a tree partially concealed in foliage canopy and not silhouetted against the sky. From here it keeps a sharp look-out for junglefowl, hares and other small animals coming out in the open, swooping on them with a terrific rush, striking them down and bearing them away in its talons. It has a loud high-pitched cry: *ki-ki-ki-ki-ki-ki-ki-ki ki-kee*, beginning short, rising in crescendo and ending in a scream." Distribution: practically the whole of India, Bangladesh, Sri Lanka, Myanmar. There are several races and closely allied species, differing in size and other details. The typical race is distributed in India south of the Indo-Gangetic plain, and a smaller race *S. c. ceylonensis* is found in Sri Lanka.

ARTICLE AND ILLUSTRATION
IN MEMORY OF SHRI SHANTILAL D. KOTHARI & DR. MANGALJI L. PATWA,
OF PALANPUR AND DEESA, FROM DR. AJAY P. KOTHARI, PRESIDENT,
ASTROX CORPORATION, COLLEGE PARK, MARYLAND, USA

Changeable Hawk Eagle (Crested Hawk Eagle) *Spizaetus cirrhatus* (Gmelin)
*Birds of* Asia, Vol. III, Parts XIII–XVIII, by John Gould, 1861–66. Painted by J. Wolf & Henry C. Richter.

In Memory of Kekoo Naoroji, from Rishad Naoroji

# An Adventure with a Cobra de Capella

"Miles"

I had escaped for a day from the incessant routine of military duties, for which the Potsdam of India is so justly celebrated. It was about the conclusion of the monsoon of 1835; the quail were abundant, and after some hours of hard fagging, through dark and heavy grass, I felt inclined to rest; an adjacent Tamarind tree of noble growth yielded an inviting shelter, from a sun that for the season of the year, was oppressively hot. The few beaters who had accompanied me had set off to a neighbouring Gaum (village) to obtain some refreshment. Left to myself I was much to my own satisfaction employed in counting over the contents of a well filled game bag, and mentally *portioning off lots*, as presents to my different friends. From this state of pleasing indolence, which a *shooter* is apt to indulge in after severe fatigue, I was aroused by the furious barking of my dogs; on turning around I beheld a snake of the Cobra de Capella species, directing its course to a point that would approximate very close upon my position; in an instant I was on my feet. The instant the reptile became aware of my presence, in nautical phraseology, it *boldly brought to*, with expanded hood, eyes sparkling, neck beautifully arched; the head raised nearly two feet from the ground, and oscillating from side to side, in a manner plainly indicative of a resentful foe. I seized the nearest weapon of my wrath, a short bamboo, left by one of the beaters, and hurled it at my opponent's head. I was fortunate enough to hit it beneath the eye. The reptile immediately fell from its imposing attitude, and lay apparently lifeless. Without a moment's reflection I seized it a little below the head, hauled it beneath the shelter of the tree, and very coolly sat down to examine the mouth of the poisoned fangs of which naturalists speak so much. While in the act of forcing the mouth open with a stalk, I felt the head sliding through my hand, and to my utter astonishment became aware that I now had to contend against the most deadly of reptiles in its full strength and vigour. Indeed I was in a moment convinced of it, for as I tightened my hold of the throat, its body became wreathed round my neck and arm. I had raised myself from a sitting posture to one knee, my right arm (to enable me to exert my strength) was extended; I must, in such an attitude, have appeared horrified enough to represent a deity in the Hindu mythology, such as we so often see emblazoned on the portals of their temples. It now became a matter of self-defence: to retain any hold it required my utmost strength to prevent the head from escaping, as my neck became a PURCHASE, for the animal to pull upon. If the reader is aware of the universal dread in which the *Cobra de Capella* is held throughout India, and the almost instant death, which invariably follows its bite, he will, in some degree, be able to imagine what my feelings were at the moment:– a shudder, a faint kind of disgusting sickness pervaded my whole frame, as I felt the cold, clammy hold of the reptile tightening round my neck. To attempt any delineation of my sensations would be absurd and futile: let it suffice they were most horrible, I had now almost resolved to resign my hold. Had I done so this tale would never have been written; as no doubt the head

**Eurasian Eagle Owl >**
***Bubo bubo bengalensis***

A very large owl, mottled tawny-buff and blackish brown with conspicuous upright ear-tufts above large orange eyes. Nocturnal, sits motionless during the day amongst rocks and ravines and occasionally in foliage of the trees. Call: a deep, resonant, hollow *bu-bo* (2nd syllable prolonged) repeated at intervals; not loud but with a curious penetrating and far-carrying quality. Resident, throughout much of the subcontinent, except parts of the northwest and northeast and Sri Lanka. This is the commonest of the large owls of India, being very abundant in northern and central India.

Eurasian Eagle Owl (Great Horned Owl) *Bubo bubo bengalensis* (Franklin)
*A Century of Birds from the Himalaya Mountains,* by John Gould, 1832. Painted by Elizabeth Gould.

Courtesy Poultry Development Promotion Council

Oriental Bay Owl (Bay Owl) *Phodilus badius* (Horsfield)
*Birds of Asia*, Vol. IV, Parts XIX–XXIV, 1867–72. Painted by John Gould & Henry C. Richter.

Courtesy Pitti Laminations Limited

**Rikki-tikki-tavi confronts Nag**. Drawing by W.H. Drake. *The Jungle Book* by Rudyard Kipling, with illustrations by J. Lockwood Kipling, W.H. Drake, and P. Frezeny, 1895.

### < Oriental Bay Owl
*Phodilus badius*

A small chestnut bay owl with vinous pink facial disc, white ruff tipped with chestnut and black, and short ear-like tufts projecting above sides of the head. Above, chestnut, spotted with black and buff. Tail chestnut, barred with black. Below, white necklace and further down vinous pink spotted with black and white. Short rounded wings, rapid flight action and distinctive head shape helps in identification. Very little is known about its habits as the bird is strictly nocturnal and seldom seen. It is helpless in daylight; spends daytime lurking in dark holes and hollows in tree-trunks. Recorded in the Teesta valley up to 1,500 m. It is confined in the Sikkim Himalayas to heavy foothills forest (Sálim Ali, *Birds of Sikkim*). Also found in Nepal and other parts of NE India.

would have been brought to the extreme circumvolution to inflict its deadly wound. Even in the agony of such a moment I could picture to myself the fierce glowing of the eyes, and the intimidating expansion of the hood, 'ere it fastened its venomous and fatal hold upon my face or neck. To hold it much longer would be impossible.

Immediately beneath my grasp there was an inward working, and creeping of the skin, which seemed to be assisted by the very firmness with which I held it – my hand was gloved. Finding in defiance of all my efforts that my hand was each instant forced closer to my face, I was anxiously considering how to act in this horrible dilemma, when an idea struck me that, was it in my power to transfix the mouth, with some sharp instrument, it would prevent the reptile from using its fangs, should it escape my hold of it. My gun lay at my feet, the ramrod appeared the very thing required, which with some difficulty I succeeded in drawing out, having only one hand disengaged. My right arm was now trembling from over exertion, my hold becoming less firm, when I happily succeeded in passing the rod through the lower jaw up to its centre. It was not without considerable hesitation that I suddenly let go my hold of the throat, and seized the rod in both hands; at the same time bringing them over my head with a sudden jerk, disengaged the hold from my neck, which had latterly become tight enough to produce strangulation. There was then little difficulty, in freeing my right arm, and ultimately to throw the reptile from me to the earth, where it continued to twist and writhe itself into a thousand contortions of rage and agony. To run to a neighbouring stream to have my neck, hands, and face, in its cooling waters, was my first act, after despatching my formidable enemy.

Thus concludes a true, though plainly told tale, as a moral, it may prove – that when man is possessed of determination, coolness, and energy, combined with reason, he will generally come off triumphant, though he may have to circumvent the subtlety of the snake, or combat the ferocity of the tiger.

Kirkee, near Poonah.

From *The Bengal Sporting Magazine*, Vol. VIII, 1836.

# The Poisonous Snakes of the Bombay Presidency

H.M. Phipson, c.m.z.s., Hon. Sec.

A fortnight ago one of our local newspapers stated that there were not more than three, or perhaps four, poisonous snakes in the Bombay Presidency. I felt that we ought not to allow such a statement to pass unchallenged, especially as our own collection furnished evidence that nine poisonous snakes, at least, are to be found in the Presidency, and that, according to the greatest authority on the subject, Dr. Gunther, a tenth, which we have not as yet obtained, is an inhabitant of the Deccan. I consequently gave the *Times of India* a list of the poisonous snakes in our possession, all of which had been killed in this Presidency, a list which, I think, reflects great credit on this Society, when the short time during which the collection has been got together is taken into consideration. Some of the measurements we were able to give have already attracted the notice of the press in other parts of India, and I therefore think it would be of interest to the members present, if I were to draw their attention to the specimens we possess of these particular snakes. We have, you will observe, specimens of the following poisonous snakes, all of which were killed in this Presidency:–

Colubrine.– 1. *Ophiophagus hannah*. 2. *Naja naja*. 3. *Bungarus caeruleus*. 4. *Calliophis melanurus*. 5. *Calliophis nigrescens*.

Viperine.– 6. *Daboia russelii*. 7. *Echis carinatus*. 8. *Trimeresurus malabaricus*. 9. *Hypnale hypnale*.

1. We will take, first, the great Colubrine snake, the *Ophiophagus hannah*, the "Hamadryad" or "King Cobra," which is probably the largest poisonous snake in the world. I say probably, as there is one in New Guinea, *Lachesis mutus*, a viperine snake belonging to the Crotalidae, which is said to reach 14 feet in length. Fortunately, the Hamadryad is not very common. Dr. Gunther, the well-known ophiologist, says that the Hamadryad is found in all parts of the Indian Continent, in the Andamans (where, I hear, it is eaten by the natives), the Philippines, Java, Sumatra and Borneo. As its name implies, it feeds principally on snakes and other reptiles. Owing to the fact of its expanding a "hood" it is frequently mistaken for a cobra, but, as you will see by comparing the specimens the plates or shields on the head of the Hamadryad differ materially from those of the cobra. According to Sir Joseph Fayrer, the natives of Bengal call it the "Sunkerchor," a "breaker of shells," but he gives no explanation of this name. The snake-men about here do not appear to know the Hamadryad, but it is, undoubtedly, an inhabitant of this Presidency. We have received a skin of one from Carwar measuring 12 feet 6 inches, and another from the Goanese Ghauts which is 15 feet 5 inches in length. Major Beddome of Madras, says he has killed one nearly 14 feet near Cuttack in Bengal, where it is common. A few years ago one was caught

in the Konkan by Mr. Bulkley, who tried to take it to England alive so we have ample proof of its occurring in this part of India.

2. *Naja naja* (Linn.), the Cobra, is too well known to need description. It is found all over India up to 8,000 feet in the Himalayas. There are a great number of varieties, differing in colour and markings, many of which are figured in Sir Joseph Fayrer's Thanatophidia of India. The natives, who give separate names to these varieties, maintain that they are distinct species, and that they differ considerably, not only in appearance, but in their habits. The cobra lays from twelve to twenty eggs, once a year, during the rains, and the young show signs of their venomous power at a very early stage. Those hatched in this Society's rooms last year killed a small Malay Python (*P. reticulatus*), which was placed in their cage a few days after they were born. They attacked it at once, biting it viciously across the back. The Python showed great signs of fear, but made no attempt at retaliation. It was at once removed to another cage, but died in about twelve hours. We have many specimens of the cobra in our collection, amongst which is a young one preserved in the act of emerging from its egg. In this specimen, the foetal tooth with which the young snake cuts its way out of the strong parchment-like egg, can be clearly seen with a magnifying glass. This foetal tooth is shed as soon as it has served its purpose, and is, in fact, expelled the first time the snake darts out its tongue, which it usually does directly its head appears from the egg. Some of these little cobras thrived for several months on young lizards, but the others would not feed and died in about two months. They measured 7½ inches when born, and were very fat. At the end of the two months they had lost all their plumpness, but had increased their length by nearly 1½ inches. It is very extraordinary that the original nourishment obtained from the egg should be capable of sustaining them for so long a period. The cobra is an exceedingly *timid* snake, but it can be easily tamed with kindness, as you know from the living specimen in the Society's rooms. It is worthy of note that the cobra is about the only poisonous snake which those arrant impostors, the so-called "snake-charmers", ever have anything to do with. I never lose an opportunity of fraternizing with these gentlemen in the hope of obtaining specimens we are in want of, but on no occasion have I ever seen any other poisonous snake in their baskets except the cobra. The explanation of this lies, I believe, in the fact that the cobra is the only poisonous snake which can be easily and safely handled. You have only to attract its attention with one hand, while you seize it in the middle of the body with the other, and the snake is yours. It strikes in every direction, *especially at any moving object,* but it never seems to occur to it to turn and bite the hand that is holding it, as almost all other snakes would do at once. The snake-charmers have from time immemorial made great capital out of the knowledge of this simple fact. Their performances with the cobra are known to you all. The snake is taken from the basket, when a slight slap across the back brings it at once into its striking posture. *It is the constant movement of the musical instrument in front of the snake that keeps it erect, and not the noise produced.* Snakes have no external ears, and it is very doubtful whether the cobra hears the music at all. The vipers, which are far less timid, cannot be frightened in this manner, and consequently they are not used for these performances. The snake-men will tell you that the Daboia, the largest viper, or adder, of the East, is a dull snake with no ear for music, and it is interesting to note that they have evidently been repeating this nonsense ever since the time of David – *vide* Psalms LVIII – "like the death adder that stoppeth her ear; which will not harken to the voice of charmers, charming never so wisely."

The cobras in the Society's rooms feed freely on young rats, birds and toads.

3. We next come to the Krait (*Bungarus caeruleus*), which is also a very well-known snake. It is exceedingly poisonous, and is common in nearly all parts of India. We have a number of specimens in our collection from the Bombay Presidency and from Bombay itself. I have lately received two from Malabar Hill. The one contained a

**"Cobra de Capella"**. *Routledge's Picture Natural History* by the Rev. J.G. Wood, engraved by the Dalziel brothers, 1885.

"brown tree snake" (*Boiga gokool*), and the other a Dhaman (*Ptyas mucosus*), so that we have good evidence of its snake-eating propensities. The dark variety of the common and harmless common wolf snake (*Lycodon aulicus*) is, you will observe, very like the Krait in outward appearance, but you can readily distinguish the Krait by the large hexagonal scales down the centre of the back. The Burmese Krait or Banded Krait (*Bungarus fasciatus*), of which we have several beautiful specimens, is not found, I believe, in any part of this Presidency, although it occurs in parts of Bengal and Lower India.

4. Our fourth poisonous Colubrine land snake is the slender coral snake (*Calliophis melanurus*) which does not possess any popular name that I am aware of. It is a ground snake, and lives chiefly on other small snakes. Dr. Gunther says that the Calamariae, which they much resemble in appearance, are their principal food. This snake, although so small, is undoubtedly poisonous. We have two specimens, one from the Konkan and the other from Bandora (Bandra).

5. I have just received a telegram from Mr. G.W. Vidal, C.S. to the effect that the specimen of the common Indian Coral Snake or Gunther's Coral Snake (*Calliophis nigrescens*) which he deposited some time ago with the Society, was found by him in Carwar, thus adding another poisonous snake to the list of those found in this Presidency. The upper parts of this snake are black and the lower uniform red. It grows to about four feet in length.

6. We now come to the Viperine snakes, first and foremost of which is the deadly Russell's Viper *Daboia russelii*, the Gunus (Ghonus) of the natives, known to Europeans in India as the Chain Viper and in Ceylon as the Tic Polonga. It is common in the Island of Bombay, and is, I believe, found in most parts of the Presidency. According to Sir Joseph Fayrer's experiments, the poison of this snake, although very different in its action, is almost, if not quite, as fatal as that of the cobra. It has, as you will observe, exceedingly long fangs and a good supply of spare ones behind ready to take the place of those in front should they be broken. From its sluggish habits, its fierceness, and the great length of its fangs, it is to be dreaded, I think, more than any other snake in this country. Most of the authorities give 50 inches as its length, but we have the head of one killed by by Mr. J.C. Anderson, in Hurda, Central Provinces, which was 61½ inches. Judging from the size of the head, and the evidence of the piece of string with which the snake was measured, there is little doubt that the correct length has been stated. Like most of the vipers it is difficult to keep in confinement, but it is very tenacious of life, and has been known to live for a whole year without food. It is an exceedingly handsome snake, especially when young.

7. The only other true viper in this country is the Saw-scaled Viper *Echis carinatus*, known here as the Phoorsa and in Sind as the Kupper. We have received it from many parts of the Presidency, and in some districts – Rutnagherry for instance – it is found in great numbers. I have never heard of its being killed in the Island of Bombay, although the harmless "brown tree snake" (*Boiga gokool*), which somewhat resembles it, is often sent to me as a Phoorsa. You will readily distinguish them, as the head of the Echis, like all vipers, is covered with scales, whereas that of the *Boiga gokool* has plates or shields. Dr. Gunther was, when he issued his book on the Indian Reptiles, under the impression that the bite of this little viper was not absolutely fatal, but it has since been proved that in certain districts the mortality from the Phoorsa is very great.

8. The Green Viper (*Trimeresurus malabaricus*) belongs to the family of Crotalidae, or Pit Vipers, so called from a curious pit or cavity between the nostril and the eye, the use of which is not known [see Editors' Note 1]. The dreaded rattle-snake of America belongs to the same family. There are eight species of Trimeresuri in India,

**Rufous-vented Laughingthrush >**
*Garrulax gularis*
A uncrested laughingthrush the size of a myna, chiefly rufous olive-brown, dark grey, and yellow. Above, crown and nape slaty grey; olive-brown tinged rufous; tail largely rufous. Below, throat primrose-yellow; sides of breast dark grey, legs and feet orange-yellow. Affects dense evergreen undergrowth in bamboo and scrub jungle. Gregarious. A great skulker, difficult to observe. Feeds on the ground, rummaging among the mulch. Call characteristic of the laughingthrushes: shrill squeaks, chattering and cackling, often in discordant choruses of "laughter". Resident, foothills of eastern Bhutan and Arunachal Pradesh, and south through the hills of NE India (Assam, Meghalaya, Mizoram, Nagaland) and Bangladesh between 1,000 and 1,800 m. Once in abundance when Dr. Sálim Ali surveyed the area a few decades ago but very few recent records.

Rufous-vented Laughingthrush *Garrulax gularis* (Jerdon)
*Birds of Asia*, Vol. IV, Parts XIX–XXIV, by John Gould, 1867–72. Painted by John Gould & Henry C. Richter.

but we have, at present, in our collection, only *T. malabaricus* from the Bombay Presidency. It appears to be common on the Ghauts, as we receive many from Khandalla, Egutpuri and Mahableshwar. Dr. Gunther states that another species, the Horse-Shoe Viper, *T. strigatus*, is found in the Deccan, and I hope before long some of our up-country members will be able to send us one in order that we may have specimens of the ten poisonous snakes, which are now known to belong to this Presidency. (Editor's note: A specimen has since been received from Mr. H.S. Wise, which was killed in Carwar.) It is just possible that an eleventh, *Peltopelor macrolepis*, may also occur in the Canarese jungles, as it is said to be common a little further south.

9. We now come to the Hump-nosed Pit Viper (*Hypnale hypnale* Merrem) or the Carawala, which was found in Carwar by Mr. G.W. Vidal, C.S. Its head-quarters are in Ceylon, where it is greatly dreaded, but, like so many of the Ceylon fauna, this snake is to be found along the Malabar Coast, but probably not further north than Carwar.

I have to-day only dealt with the poisonous land snakes of this Presidency, but all the true sea-snakes are, as you know, poisonous.

I may state that we have at present in our collection specimens of the following species [see Editors' Note 2]:–
*Hydrophis torquatus diadema.* (Gunther.) [This is now *H. obscurus* Daudin.]
*Hydrophis robusta.* (Gunther.)
*Lapemis curtus.* (Gunther [Shaw].)
*Hydrophis aurifasciatus.* (Murray.)
*Hydrophis Phipsoni.* (Murray.)
*Hydrophis Guntheri.* (Murray.)

**Chestnut-bellied Rock Thrush >**
*Monticola rufiventris*
Male brilliant blue above with blackish mantle, eye- and ear-patches and sides of neck black. Below, throat blackish blue; rest chestnut. Female olive-brown above with dark crescent-shaped band. Below, squamated dark brown and buff. Distinguished from similar female Rock Thrush by larger size, orange-buff eye-ring and whitish throat-patch. Usually seen perched high up in tall forest trees, jerking its tail up and down. Forages mainly on the ground, in search of insects. Also eats berries and other forest fruits. Call: a harsh rattle *chhrrr*. Song, a pleasant warbling song, *teetatewleedee-tweet tew*, repeated several times (Sálim Ali, *Birds of the Eastern Himalayas*). Resident, common in open coniferous and broad leaved forest, subject to vertical movements. Himalayas from Murree in Pakistan east to Arunachal Pradesh and NE India. Breeds between 1,200 and 3,300 m; winters down through foothills and duars.

**"Dancing Snake and the Musicians"**. From a drawing taken on the spot by Baron de Montalembert, 1807. *Oriental Memoirs*, Vol. I, by James Forbes, 1812.

Chestnut-bellied Rock Thrush *Monticola rufiventris* (Jardine & Selby)
*Birds of Asia*, Vol. III, Parts XIII–XVIII, by John Gould, 1861–66. Painted by John Gould & Henry C. Richter.

Courtesy Chhotalal Premchand Shah, Diamondstar

*Hydrophis Lindsayi.* (Gray.)
*Hydrophis chloris.* (Daud.)
*Entrydrina bengalensis.* (Gray.)
*Pelamis bicolor.* (Daud.) [This is now *Pelamis platurus* (Linn.), the Flat-Tail or the Black and Yellow Sea Snake.]

Read at the Society's meeting on 5th September 1887, published in *JBNHS*, Vol. II.

**HERBERT MUSGRAVE PHIPSON** (1850–1936) came to India in 1878, established the firm of Phipson & Co., Wine Merchants in 1883, and left India in 1906. He was in England when the eight original founders of the Bombay Natural History Society met at the Victoria & Albert Museum, Bombay (now the Bhau Daji Lad Museum) on 15th September 1883. He joined the Society on his return from England in the same year and in January 1884 offered a room in his office at 18, Forbes Street as a more central place for the Society's meetings and for its collections. When the Society needed a bigger place Phipson again provided a solution by offering part of the larger premises he had acquired at 6, Apollo Street (now Shaheed Bhagat Singh Road) where the Society remained till 1958. His successors in business took over his post of Honorary Secretary and Editor of the Journal and each in their time made their contribution to the Society's progress. From March 1886, when he took over the office of Honorary Secretary from E.H. Aitken (EHA) to April 1906 when he left India, Phipson was the heart and soul of the Society. Through these twenty years he edited the Society's Journal, for a year in collaboration with Robert Sterndale, then as a sole editor for fifteen years, and finally jointly with W.S. Millard, his immediate successor in office. He did not succeed in his dream of establishing a zoological garden conducted and managed by the Society because the Municipality was unwilling to allow the use of the site selected by Phipson for the purpose. In his ambition to provide the city with a fine natural history museum Phipson was more successful; the admirable Natural History Section of the Prince of Wales Museum (now Chhatrapati Shivaji Maharaj Vastu Sangrahalaya) is largely the fruit of his initiative and exertions. Phipson's name as a naturalist is fittingly commemorated by zoologists naming several new discoveries after him, e.g. the sea snake *Hydrophis phipsoni* [*Hydrophis cyanocinctus*], the earth snake *Silybura* [*Uropeltis*] *phipsoni*, the scorpion *Isometrus phipsoni*, the whip scorpion *Phrynicus phipsoni*, and the galeod spider *Rhagodes phipsoni*. The beautiful flying squirrel *Petinomys phipsoni* [*Vorder manni*], though discovered by the Society's Mammal Survey long after Phipson's departure from India, was also named in his honour to perpetuate the memory of his dedicated association with the Society and with Indian Natural History. – From *JBNHS*, Diamond Jubilee Issue, Vol. LXXV, December 1978.

**EDITORS' NOTES**
1. It is now known that the pit is sensitive to infrared radiation which helps the snake locate its prey in the dark.
2. It was common practice among naturalists of the 19th and early 20th centuries to create new species based on minor variations in morphology. Later, many of these were found not to be valid species, as the characters were not considered sufficiently different, hence these were merged with existing species. Thus, except for *Hydrophis diadema* (now *H. obscurus*), *Lapemis curtus*, and *Pelamis bicolor* (now *P. platurus*), all the sea-snakes listed by Phipson are invalid species.
3. The Bombay Presidency covered a vast area from Sind and Kathiawar to Canara including Kutch, Palanpur, Ahmedabad, Mahikantha, Revakantha, Surat, Dangs, Colaba, Ratnagiri, Savantwadi, Poona, Sholapur, Kolhapur, Khandesh, Ahmadnagar, Belgaum, and Dharwar districts.

– A.S.K. & B.F.C.

**Yellow-throated Laughingthrush >**
*Garrulax galbanus*

A small laughingthrush with yellow under-parts, black mask, grey crown and nape, greyish olive flanks, pale olive-brown upper-parts, and greyish tail with black tip. Usually found in pairs or in small parties, sometimes in larger flocks. Found in open jungle and tall grass intermixed with trees and shrubs in wet evergreen, subtropical biotope, between c. 800 and 1,800 m. Feeds on the ground, flying into trees when disturbed. Resident, southern Assam (Cachar), Nagaland, Manipur, Mizoram, and Chittagong hill tracts in Bangladesh.

Yellow-throated Laughingthrush *Garrulax galbanus* Godwin-Austin
*Birds of Asia*, Vol. V, Parts XXV–XXX, by John Gould, 1873–77. Painted by John Gould & William Hart.

Courtesy Dimexon Diamonds Limited

# A Giant Squid Attacking a Ship

A SUCCESSOR TO THE SEA SERPENT – [The following was published in the column devoted to shipping news] – "We had left Colombo in the steamer *Strathowen*, had rounded Galle, and were well in the bay, with our course laid for Madras, steaming over a calm and tranquil sea. About an hour before sunset on the 10th of May we saw on our starboard beam and about two miles off a small schooner lying becalmed. There was nothing in her appearance or position to excite remark, but as we came up with her I lazily examined her with my binocular, and then noticed between us, but nearer her, a long, low, swelling lying on the sea, which from its colour and shape I took to be a bank of seaweed. As I watched, the mass, hitherto at rest on the quiet sea, was set in motion. It struck the schooner, which visibly reeled, and then righted. Immediately afterwards the masts swayed sideways, and with my glass I could clearly discern the enormous mass and the hull of the schooner coalescing – I can think of no other term. Judging from their exclamations, the other gazers must have witnessed the same appearance. Almost immediately after the collision and coalescence the schooner's masts swayed towards us, lower and lower; the vessel was on her beam-ends, lay there a few seconds, and disappeared, the masts righting as she sank, and the main exhibiting a reversed ensign struggling towards its peak. A cry of horror rose from the lookers-on, and, as if by instinct, our ship's head was at once turned towards the scene, which was now marked by the forms of those battling for life – the sole survivors of the pretty little schooner which only 20 minutes before floated bravely on the smooth sea. As soon as the poor fellows were able to tell their story they astounded us with the assertion that their vessel had been submerged by a gigantic cuttlefish or calamari, the animal which, in a smaller form, attracts so much attention in the Brighton Aquarium as the octopus. Each narrator had his version of the story, but in the main all the narratives tallied so remarkably as to leave no doubt of the fact. As soon as he was at leisure, I prevailed on the skipper to give me his written account of the disaster, and I have now much pleasure in sending you a copy of his narrative:– 'I was lately the skipper of the *Pearl* schooner, 150 tons, as tight a little craft as ever sailed the seas, with a crew of six men. We were bound from the Mauritius for Rangoon in ballast to return with paddy, and had put in at Galle for water. Three days out we fell becalmed in the bay (lat. 8°50' N., long. 8°45' E.). [This is about 600 miles from Car Nicobar where Bullen saw the fight between a giant squid and a sperm whale.] On the 10th of May, about 5 p.m., – eight bells I know had gone, – we sighted a two-masted screw on our port quarter, about five or six miles off; very soon after, as we lay motionless, a great mass rose slowly out of the sea about half a mile off on our larboard side, and remained spread out, as it were, and stationary; it looked like the back of a huge whale, but it sloped less, and was of a brownish colour; even at that distance it seemed much longer than our craft, and it seemed to be basking in the sun. "What's that?" I sung out to the mate. "Blest if I knows; barring its size, colour, and shape, it might be a whale," replied Tom Scott; "and it ain't the sea sarpent," said one of the crew, "for he's too round for that ere crittur." I went into the cabin for my rifle, and as I was preparing to fire, Bill Darling, a Newfoundlander, came on deck, and, looking at the monster, exclaimed, putting up his hand, "Have a

**Red-faced Liocichla >**
*Liocichla phoenicea*
A striking olive-brown laughingthrush with sides of head bright crimson and extensive crimson on the wings. Tail black with reddish tip and outer features. Sexes alike. Found in pairs in breeding season, otherwise in small parties. Very skulking and difficult to observe. Call: squeaky conversational notes and also loud plaintive song of 5 or 6 notes, the last 3 or 4 on the same note. Resident, locally common in dense undergrowth in evergreen or moist deciduous forest between 900 and 1,800 m in Himalayas from Nepal east to Arunachal Pradesh, NE India and Bangladesh. Comes down to duars in winter.

Red-faced Liocichla (Crimson-winged Laughingthrush) *Liocichla phoenicea* (Gould)
*Birds of Asia*, Vol. IV, Parts XIX–XXIV, by John Gould, 1867–72. Painted by John Gould & Henry C. Richter.

Courtesy UTI Mutual Fund

care, master; that ere is a squid, and will capsize us if you hurt him." Smiling at the idea, I let fly and hit him, and with that he shook; there was a great ripple all round him, and he began to move. "Out with all your axes and knives," shouted Bill, "and cut at any part of him that comes aboard; look alive, and Lord help us!" Not aware of the danger, and never having seen or heard of such a monster, I gave no orders, and it was no use touching the helm or ropes to get out of the way. By this time three of the crew, Bill included, had found axes, and one a rusty cutlass, and all were looking over the ship's side at the advancing monster. We could now see a huge oblong mass moving by jerks just under the surface of the water, and an enormous train following; the oblong body was at least half the size of our vessel in length and just as thick; the wake or train might have been 100 feet long. In the time that I have taken to write this the brute struck us, and the ship quivered under the thud; in another moment, monstrous arms like trees seized the vessel and she heeled over; in another second the monster was aboard, squeezed in between the two masts, Bill screaming "Slash for your lives;" but all our slashing was of no avail, for the brute, holding on by his arms, slipped his vast body overboard, and pulled the vessel down with him on her beam-ends; we were thrown into the water at once, and just as I went over I caught sight of one of the crew, either Bill or Tom Fielding, squashed up between the masts and one of those awful arms; for a few seconds our ship lay on her beam-ends, then filled and went down; another of the crew must have been sucked down, for you only picked up five; the rest you know. I can't tell who ran up the ensign. – JAMES FLOYD, late master, schooner *Pearl*.'" – *Homeward Mail*. [This means the story was sent to London by a homeward-bound mail vessel.]

I have tried hard but unsuccessfully to find confirmation of this incident – from Lloyd's, the National Maritime Museum, the General Register of Shipping and

Seamen, shipping lines and other likely sources. The reliability of the account must, therefore, be judged, by internal evidence alone.

The opening of *The Times*' report indicates that the story was well-known in India. Where did it come from? Men and ships are named, date and time are given, the position is pin-pointed to minutes of latitude and longitude, and circumstantial accounts of the incident are recorded from both the onlookers' and the victims' points of view. If it were all fiction – and there seems no alternative to a complete hoax if the story is untrue – how did the hoaxer persuade newspapers to publish the baseless story? Evidently the Editor of *The Times* was satisfied it was not a hoax.

I sent a copy of *The Times*' report to Commander Groenningsaeter, the only man I know who has witnessed such an incident, and he replied that the "method of attack" on the *Pearl* seemed to be identical to that on his own ship. He suggests that the *Pearl* may "have had a very low stability being in ballast and with sails up. Its ballast may have shifted."

Groenningsaeter's suggestion may explain why the squid, although so much lighter than the *Pearl* and without having a solid object to give it purchase, was able to capsize it. These two facts, the great disparity in weight and lack of support, are probably the strongest arguments against the authenticity of the report.

*The Times*' account shows evidence of being written by eye-witnesses. The appearance of the animal as it lay on the surface; the jerky movement as the squid propelled itself by jets from its funnel; the trailing arms ("the wake or train"); and the use of the word coalescing ("I can think of no other term") when the squid swarmed over the schooner – that is just the word to describe such an action – all these are exactly right, and indicate that the reports came from people who were describing what they actually saw.

To me, however, the most convincing evidence of authenticity is the casual remark that the man who warned the master of the *Pearl* not to molest the squid was a Newfoundlander. As the preceding pages have shown, at the time of this incident the one place in the world where men were most likely to know about large squids, and their ferocity if attacked, was Newfoundland.

Information travelled much more slowly then than now, and it is unlikely that Moses Harvey's accounts of two squids, which had been written only some six months before, would be common knowledge in India at the time of the alleged sinking of the *Pearl*. Moreover, at that time, May 1874, nobody knew that Newfoundland was to become famous during the next few years for its stranded giant squids. The most *reasonable* explanation seems to be that the account was a report of an actual incident, including the presence on the *Pearl* of a man from the one place where, at that time, giant squids and their behaviour were reasonably well-known.

From *The Times*, London, 4th July 1874.

ARTICLE AND ILLUSTRATION COURTESY VASANT J. SHETH MEMORIAL FOUNDATION

# Wildlife around Cambay and Ahmedabad

## During the eighteenth and nineteenth centuries when Tigers used to enter the city limits of Ahmedabad and Lions were found near Cambay

JAMES CAMPBELL

About a hundred years ago tigers, lions, and other large game were common in Ahmedabad. Tigers (1783) were found in the desolate ground outside of the city walls, and in the Dholka subdivision dense forests near the Sabarmati were the resort of lions and tigers. Forbes in his Oriental Memoirs has preserved Sir Charles Malet's account of a lion hunt in those forests in the year 1780. Near Kura, about thirty miles north of Cambay, a place of impenetrable woods, not far from the Sabarmati, the traces of some large animals of the tiger class were found. Failing to beat them out, goats were tied to trees and marksmen set over them. About midnight four large animals came near one of the trees, and two of them trying to carry off the goats were wounded. Next day with a large body of beaters they were tracked through a forest, stretching for miles so thick that the sportsmen had to force their way on hands and knees. The wounded animals when sighted were found to be lions. They made their way into a still closer thicket, and were forced out only by the device of collecting and driving into their lair a herd of buffaloes. When they moved out one of them was killed. The people called it the camel tiger, *untia vagh*, the strongest and fiercest of the race. In colour it was rather yellower than a camel without spots or stripes, "not high, but powerfully massive with a head and forepart of admirable size and strength." Oil was extracted and the flesh eaten by the Vaghris. A few years later (1787) in the same part of the district tigers were met in the high grass fields, and as late as 1824 the salt flats between Dholka and the Sabarmati, covered with a thick growth of marsh shrubs, were infested with both lions and tigers. About the same time (1825) in Modasa and the other eastern districts, especially on the river banks, tigers were numerous, doing much harm to cattle but little to men. Close to Ahmedabad the Shahi Bag and other old gardens were infested with tigers, and as late as 1840 one was shot in the Queen's Mosque in Mirzapur. As in other parts of Gujarat the increase of population and the spread of tillage have, during the last fifty years, done much to drive off the larger class of game.

The Tiger, *vagh*, *Panthera tigris*, is now (1877) almost never found within Ahmedabad limits. In the east of Modasa from one to five tigers are generally killed every season. But the tiger's haunts and the actual shooting are generally a few miles over the Mahi Kantha border. The Panther, *dipdo*, *Panthera pardus*, is found in Modasa

**White-breasted Waterhen >**
*Amaurornis phoenicurus*
A familiar slaty-grey stub-tailed, long-legged marsh bird with prominent white mask and breast and chestnut under-parts. Bill olive-green with a red patch at the base of the upper mandible; legs greenish yellow. Sexes alike but female is slightly smaller. This is one of the commonest waterbirds of India. Affects moist ground overgrown with tangles of bushes, pandanus brakes, etc, on the margins of jheels and ponds. The stumpy tail, carried erect as the bird stalks or skulks along, is constantly jerked up, flaring the chestnut below into prominence. The species is remarkable for its calls, being extremely noisy. The ordinary note is a sharp metallic sound, much like that of a pestle and mortar, heard chiefly on cloudy overcast days and often all through the night. Distribution: throughout India (including Andaman and Nicobar Islands) up to the base of the Himalayas; Bangladesh, Pakistan, Sri Lanka, Maldives, Myanmar.

White-breasted Waterhen *Amaurornis phoenicurus* (Pennant)
*Birds of Asia*, Vol. IV, Parts XIX–XXIV, by John Gould, 1867–72. Painted by John Gould & Henry C. Richter.

Courtesy Aditya Realtors Limited

and sometimes in large tracts of grass and brushwood, *bir*, in different parts of the district. Of panthers three each year were killed in 1873 and 1874, and five each year in 1876 and 1877. The Black Bear (Sloth Bear), *rinchh*, *Melursus ursinus*, is almost unknown. It is sometimes found in Modasa strayed from the Idar forests. The Wolf, *varu*, *Canis lupus*, is common in the west of the district on the low-lying salt lands near the Nal. The Hyaena, *taras*, or *tarak*, *Hyaena hyaena*, found wherever there are hills and brushwood, is commonest in Gogha, Parantij, and Modasa. The Jackal, *sial*, *Canis aureus*, and the Fox, *lokdi*, *Vulpes bengalensis*, are common everywhere. The Wild Boar, *dukkar*, *Sus scrofa*, is found in large numbers throughout the district. Except in outlying parts the wild boar is losing his strength and fierceness. In many places four out of five have a whity brown tinge, the result of too close an intimacy with village swine.

Of the Deer tribe, the Sambar, *Cervus unicolor*, is occasionally found in Modasa. The Blue Bull, *nilgai*, *Boselaphus tragocamelus*, formerly very common, though much reduced in numbers, is still found in the plains throughout the district. The Spotted Deer, *chital*, *Axis axis*, is found only in Modasa, and is there very rare. The Antelope or Blackbuck, *kaliar*, *Antilope cervicapra*, is found in large herds throughout the district. The Indian Gazelle, *chinkara*, *Gazella bennettii*, is common in the western districts and in the rocky uplands of Parantij in the east. The Four-horned Deer, *bekri*, *Tetracerus quadricornis*, is found only in the thickly wooded Modasa ravines.

Of smaller animals, the Hare, *sasla*, *Lepus nigricollis* is found everywhere, and the

**Shah Baug**, a Summer Palace Built by the Emperor Shah Jehan on the Banks of the Sabermatty. Drawn by James Forbes, 1781. *Oriental Memoirs*, Vol. III, 1813.

The Palace stands even today and the area around it in Ahmedabad is known as Shahibag. Poet and Nobel laureate Rabindranath Tagore mentions the palace in his memoirs as he stayed there with his brother who was Chief Justice of the Gujarat High Court whose official residence it was. After the formation of Gujarat state in 1960 it was for many years the residence of the Governor of Gujarat.

Otter, *panini biladi, Lutra lutra,* in the Sabarmati and in most large sheets of water. The Indian Badger (Ratel), *adamkhor, Mellivora capensis,* during the rainy season of 1878 is believed to have done much mischief in Ahmedabad. In the former rains, reports of an evil spirit, *bhut,* were common. But as it was not accused of doing any harm, no inquiry was made. In July 1878 rumours again got abroad, and this time the evil spirit was said to have snatched a sleeping child from a house verandah, and in a very short time to have eaten it all but the head, hands, and feet. Search was made, and there was no doubt that a child had been killed and eaten. Professional trackers, called in by the police, found marks like those of a *chita* or a bear. Those they knew to be badger tracks, and traced them to a timber yard. Constables were set to watch the yard, and at night one had a shot but missed. After this, in spite of the efforts of the police, the badgers could not be traced. Meanwhile four children were carried off and eaten, one of them snatched from the mother's arms. In one case a boy thirteen or fourteen years old was attacked. But an alarm was raised and he was rescued. When the crops grew high the badgers left the city. They are known to prowl about slaughter-houses and in grave-yards to dig out dead bodies. But that the children were carried away by badgers has not yet been satisfactorily proved. The measurements of one lately shot at Bhuj, in Cutch, were, length 2⅓ to three feet, girth eighteen to twenty inches, and height fourteen to fifteen inches. The head, neck, and forequarters were very powerful.

From *Gazetteer of the Bombay Presidency,*
by James Campbell, Vol. IV, *Ahmedabad,* 1879.

NOTE: Lions, tigers, and cheetahs existed in many parts of Gujarat till the early half of the nineteenth century. Today cheetahs have vanished from India, tigers are not heard of any more in Gujarat, while lions are only found in the Gir forest.

ARTICLE AND ILLUSTRATION COURTESY TATA CHEMICALS LIMITED

# Sport and Natural History around Deesa in Northern Gujarat

Capt. C.G. Nurse, 13th Bombay Infantry

During the past three years, with the exception of some eight months spent in the Panjab during the Tirah expedition, my head-quarters have been at Deesa, and it may, perhaps, interest others if I give a short account of the sport and natural history of this delightfully sunny spot. During the months of December, January and February only is the climate pleasant; in the hot weather, *i.e.*, from March to June, and again in October and November, one is baked, and during the rains one is boiled. But, hot as it is, I personally much prefer the dry heat, which rises sometimes during the hot weather to 120°, to the moist and enervating climate of Bombay. The life has some compensations. There is small game shooting of some kind or other during most of the year, the country round is admirable for riding, and there is big game shooting to be obtained within reasonable distance.

I will begin with the Mammalia. The only monkey we see in a wild state is the common Langur, and it is not nearly so numerous here as it is near Ahmedabad, where it is a perfect nuisance. The Lion has of course long ago disappeared from this neighbourhood; the last was, I believe, killed in 1878 near the village of Bhoyen, about two miles from Deesa. The larger *Felidae* have been more than usually numerous this year; the famine has driven all the animals, both wild and domesticated, towards the streams, there being no grazing elsewhere, and Tigers and Panthers have naturally followed them. Of the smaller *Felidae*, the Jungle Cat (*Felis chaus*) is common. I have several times set a trap in the hope of getting the Indian Desert Cat (*Felis sylvestris*) but without success, as the specimens caught have always belonged to the former species. The Mungoose is, of course, abundant. I fancy there are two kinds, but I never killed one for identification.

The Wolf is common, and I have seen them within three miles of Cantonments. Jackals of course swarm, and, thanks to a sporting Colonel of Native Cavalry, who imported several foxhounds and beagles, have shown us good sport. Two kinds of Foxes occur, one with a black tip to its tail (*Vulpes bengalensis*), and one with a white tip (the Desert Fox, *Vulpes vulpes pusilla*), and both have given good runs occasionally. The Otter is fairly common; I once saw a whole family at Malana tank, about 18 miles from Deesa; there were seven or eight altogether, and they jumped into the water one after the other, so I had a good view of them. The Indian Sloth Bear of course occurs in the wooded country towards Mount Abu. Hedgehogs are common, but I fancy they all belong to one species. I once caught a specimen, and wanted to identify it. He, however, refused to unroll, so I chloroformed him and finding him

**Green Cochoa >**
*Cochoa viridis*

Male, crown to nape sky-blue above; eyebrow black; ear-patches dark blue. Green body, faintly scaled with black on mantle. Wings black with broad pale-blue band, a narrow black line across it and a small black patch. Tail blue, with outer feathers and terminal band black. Below, deep green, washed with blue on throat and belly. Tail is wholly black below. Female similar but wing feathers marked with yellowish brown instead of blue. Affects undergrowth in dense humid evergreen forests and in ravines. Solitary or in pairs. Very quiet and secretive. Feeds on ground or on trees. Food: berries, insects, molluscs etc. Call: Mild monotone whistle recorded (Boonsong and Round 1991). Resident, rare. Himalayas from N. Uttar Pradesh east to Arunachal Pradesh; NE India. Breeds between 700 and 1,500 m.

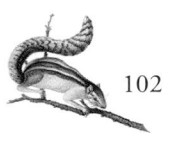

Green Cochoa *Cochoa viridis* Hodgson
*Birds of Asia*, Vol. I, Parts I–VI, by John Gould, 1850–54. Painted by John Gould & Henry C. Richter.

Courtesy Sunsoko

The Common Striped Squirrel (Palm Squirrel) *Funambulus palmarum* on a Tamarind Tree
*Oriental Memoirs*, Vol. III, by James Forbes, 1812–13. Drawn and painted in Bombay around 1779.

Courtesy Raika & Navroze Godrej

**< The Common Striped Squirrel**
*Funambulus palmarum*

T.C. Jerdon in *Mammals of India* describes this little animal thus: "Above dusky greenish-grey with three yellowish-white stripes along the whole length of back, and two fainter lines on each side; beneath whitish; tail with the hairs variegated with red and black; ears rounded. Length about 13 to 14 inches of which the tail is nearly half…. It enters houses freely, picking up crumbs, grains of rice etc., and indeed, often has its permanent abode in bungalows and out-houses, building its nest on the eaves, rafters, and in the thatch. It resorts much to the ground for its food. It usually constructs a bulky nest of grass, wool, cotton etc. Why it is named the 'palm squirrel' has puzzled the Indian naturalist, for though occasionally seen on palm trees it is so exceedingly rarely. The female has from two to four young at birth. If taken young it becomes very tame. An Indian legend runs that when Hanuman was crossing the Ganges, it was bridged over by all the animals. A small gap remained which was filled by this squirrel, and when Hanuman passed over, he placed his hand on the squirrel's back, and marks of his five fingers remained ever since on his back. When alarmed the hairs of its tail are erected at right angles, like a bottle brush."

**"An Indian Hackeree Drawn by Guzerat Oxen"**. From a drawing by Baron de Montalembert, 1807. *Oriental Memoirs*, Vol. I, by James Forbes, 1812.

to be the common *Hemiechinus micropus*, gave him his freedom as soon as he got over the effects of the anaesthetic.

Mr. Wroughton's recent interesting paper on Bats in the Society's Journal makes me wish that I had paid some attention to these interesting mammals. I only once tried to identify a specimen, which proved to be the common Yellow Bat, *Scotophilus heathi*. Flying foxes are not usually numerous, but in October and November, 1898, there seemed to be an extraordinary quantity about. I presume that some fruit must have attracted large numbers from elsewhere, as I have never seen so many anywhere as I saw then. The common Squirrel (*Funambulus palmarum*) is a perfect nuisance to anyone having a garden, as it does considerable damage by eating off young shoots. The article in the Fauna of India Series on this species seems to leave it doubtful whether it destroys birds' eggs or not, I have not the least doubt that it does so whenever it finds them unprotected, and on one occasion I purposely left an egg on the ground where it could be seen by a squirrel, with the result that it was sucked dry before my eyes in a few minutes. Rats and Mice are of course common enough, but I never tried to identify any of them. A *Gerbillus*, but whether *Tatera indica* or *Meriones hurrianae* I am not sure, is extremely common. I often wonder how they fare in the present famine year, but they seem as numerous as ever.

The Porcupine is apparently common, but owing to its nocturnal habits, seldom seen. Hares are very numerous in some places; they all seem to belong to one species, the common *Lepus nigricollis*. On one occasion two guns shot twenty-five, besides other game, in a day's sport, but this was of course exceptional. Nilgai are extremely common and very tame, but are not allowed to be shot. During the past cold weather they have always seemed in good condition, although nearly all the cattle in the country have died of starvation. I have several times seen them feeding at night in jowari fields, notwithstanding the fact that the latter are fenced all round and generally watched day and night. This is probably the reason that they appear so fat and well, as there is little or nothing for them to eat in the jungle.

Black buck are scarce in the immediate vicinity of Deesa, but Chinkara are common enough, though somewhat wild. Sambar are numerous in the hills round Abu; I am told that many of them have died of famine during the past year, and the shooting, which had become more easy than usual owing to the thinness of the jungles, has consequently been stopped. Chital I have not personally come across, but I believe that they are not uncommon in suitable localities near the foot of the hills. The "Mighty Boar" is not nearly so numerous as many of us could wish, and pigsticking, which once flourished in this neighbourhood, is at present at a somewhat low ebb.

As regards Birds, Butler's list in "Stray Feathers" of the *avifauna* of this neighbourhood is pretty exhaustive, and I made no attempt to add to it. Birds, in fact, appealed to me more as a sportsman than as a naturalist. Though I had heard of Bustard on several previous occasions, I did not come across them until the past cold weather, when I saw two and shot one. Houbara were also numerous this season, though they had hitherto been very scarce. I came across a shikari who was a perfect artist in driving both these species in whatever direction he wished. Leading a camel, and walking in a circle, he would leave the guns behind any convenient bush, and then proceed till he got to the far side of the game. He would then begin to walk in a zigzag direction towards the guns, driving the birds, which would almost invariably fly overhead well within shot. I much preferred this way of shooting them to the usual method of shooting from the back of a camel. Florican are fairly numerous at the beginning of the rains. I always have some qualms of conscience about shooting both Florican and Rain Quail at this season, but one is glad of a change from inferior mutton and bazaar *murghi*, and I fear that it is with me as with many others a case of – "Video meliora proboque, deteriora sequor."

Of Sandgrouse we usually get only two kinds in the immediate vicinity of Deesa, viz., the Common and the Painted. But during the past cold weather, the large Sandgrouse appeared in considerable numbers, and a good many were shot. Peafowl of course swarm, but are considered sacred. Jungle and Spur Fowl are pretty common at the foot of the hills. The Grey and Painted Partridge are both fairly abundant, and six kinds of Quail may be obtained, though, from the sportsman's point of view, only the Grey and Rain Quail are worth taking into consideration. The former appear in countless thousands in September; large numbers remain till the end of October, when most of them apparently go south; they reappear about the end of December, and from that time to the middle of March one can generally succeed in obtaining a fair bag. The Rain Quail appear from about the middle of June, and are plentiful enough till the end of July. After this time they are not so much *en evidence*, as the grass has become fairly high by the beginning of August, and they are thus able to breed unmolested, so far as their human enemies are concerned.

Deesa can scarcely be considered a good locality for Duck, as there is no large tank within fifteen miles. I have, however, seen or shot fifteen different kinds of Duck and Teal in the neighbourhood, chiefly during the past season. As regards Snipe, the best ground is some distance from Deesa, and here, two seasons ago, a sporting and popular doctor shot over a hundred couple to his own gun in a day. Such a day's sport is of course exceptional, but it shows what may be done under favourable circumstances.

A few Rails, which I have not taken the trouble to identify, complete the list of game birds.

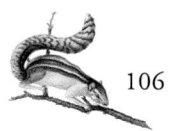

The cold weather of 1899–1900 has, owing to the famine, been an abnormal one on this side of India. Most of the usual migratory birds have scarcely appeared at all, or have come in greatly diminished numbers, and birds of prey have consequently been much fewer than usual. I have not seen a dozen Grey Quail during the whole of the cold weather. Duck and Snipe, however, which in ordinary seasons are few and far between in the immediate neighbourhood of Deesa have frequented every likely and unlikely spot in the river Banas, although the latter is in most places only a few yards wide and a few inches deep. Demoiselle Crane too, appeared in large flocks in October, flying up and down the river seeking food, but after about a month they disappeared, and I have not seen them since.

The non-migratory birds must have had a very poor time, and it appears marvellous that thousands of them have not perished of hunger. The struggle for existence during such a year as the present one must be terrible and I often wonder how some species find any food at all. Except in Cantonments, and along the bed of the river, where there is a little cultivation, there is scarcely a green leaf or a blade of grass or corn for miles, and yet every morning shortly after sunrise, and every evening about sunset enormous flocks of the common Rose-ringed Paroquet may be seen leaving or returning to the trees where they pass the night. Where do they all obtain food? There are no wild fruits, and the little grain that is being cultivated with the help of irrigation is carefully watched and guarded.

Among the Reptiles, I need hardly say that the common "Mugger" (*Crocodylus palustris*) is fairly abundant. Of the Chelonia I have only come across *Testudo elegans*, which is very numerous during the rains in the grass bhirs. The commonest house Gecko (Varied Home Gecko, Bark Gecko) appears to be *Hemidactylus leschenaulti*. On one occasion I saw one make a dash at a feather, which was blowing along the floor, mistaking it for an insect. Another time I was setting insects, and I accidentally dropped one on to the floor; before I could pick it up it was swallowed by a gecko.

Among the Lizards, a species of *Varanus* is common. Some of the natives eat its flesh, and make drumheads of its skin. The so-called "Blood-sucker" or Common Garden Lizard (*Calotes versicolor*), is extremely common, but I have scarcely ever seen one in the cold weather. I presume they hybernate in holes in the ground. The commonest Lizards are the Fan-throated Lizard, *Sitana ponticeriana*, and Hardwicke's Bloodsucker, *Brachysaura minor*; the latter may generally be seen sitting outside its holes in the evening during the hot weather.

Snakes are fairly numerous; the Cobra swarms in the grass bhirs in the rains. The Russell's Viper is common, and also the Saw-scaled Viper, *Echis carinatus*. One of the

**"A Critical Moment"**. Drawn by Robert Armitage Sterndale. *Denizens of the Jungles*, 1886.

most abundant snakes here is the Common Cat Snake, *Boiga trigonata*, which, as Boulenger says, bears an extraordinary superficial resemblance to *Echis carinatus*. I have frequently seen *Boiga trigonata* curled up on the top of cactus hedges. The elegant Leith's Sand Snake, *Psammophis leithi*, is also common, as are two, if not three, species of *Zamenis*. The plebeian-looking John's Earth Boa, *Eryx johni*, occurs, and I obtained a specimen of some species of blind snake (*Typhlops*) from beneath some rubbish in my garden.

Frogs and Toads I have not attempted to identify, and the same may be said as regards the few species of Fish that are obtainable in the neighbourhood. We get Murrel and Mahseer occasionally, but they generally have a muddy taste, and the only local Fish which is, in my opinion, worth eating is a so-called "country whitebait," but I have not the least idea to what species this belongs. About 50 species of Butterflies occur, chiefly in the rains. The most interesting are, perhaps, the various species of *Teracolus*, which is chiefly, if not exclusively, a desert genus. Some half a dozen species occur here, and some of them positively swarm during the rains. Moths are fairly numerous, but I have not yet attempted to identify those I collected, though I obtained a fair number.

One amusing incident I recall; one evening at the beginning of the rains several death's-head moths flew into the room, and settled on the ceiling. There were a great many geckos about, most of them with their abdomens considerably distended from the number of small insects they had consumed. A gecko, bolder than the rest, rushed up to one of the death's-head moths and seized it by a leg; another rushed up from the opposite side, and seized another leg. Then commenced a tug of war, which ended in the moth flying away, and both geckos falling on the floor. I could not see whether the moth got off without the loss of a leg or not, When I first arrived at Deesa, I noticed that there seemed to be more *Hymenoptera* than any other order of insects, and though I had hitherto paid little or no attention to this branch of Entomology, I determined to collect and identify as many species as possible. Bingham's volume dealing with a portion of the *Hymenoptera* in the Fauna of India Series had just been published, and the author kindly assisted me when I was in doubt, and described in this Journal some new species obtained by me. Since then I have found the study of this order of absorbing interest, and have devoted a considerable part of my spare time to the collection and identification of specimens. Even in this barren locality I have succeeded in obtaining well over 150 species, not including Ants or *Hymenoptera parasitica*. A large proportion of these are apparently new species, and have yet to be described.

The parasitic *Hymenoptera* are not numerous, except the *Evanidae*. This genus is supposed to be parasitic on *Blattidae* (Cockroaches, &c.), but I once bred a species of *Erania* from a larva of *Teracolus pleione* at Aden, so some of them are evidently parasitic on *Lepidoptera*.

I collected a fair number of *Diptera* (Flies), which I sent to England to be identified. One of the most interesting was the Horse Bot-fly, which I bred from larvae passed by a horse. I do not yet know if it is the same species that occurs in Europe.

Among the *Orthoptera* two species of Locusts are fairly common: a reddish and a yellowish kind. The latter sometimes arrives in small swarms at the beginning of the rains; but fortunately we have not had any large swarms in addition to our famine troubles. The so-called "milk-bush" (*Calotropis gigantea*), which is extremely abundant here, is frequently stripped quite bare by a species of locust, but this does not appear to be migratory, and so far as I am aware, does little or no damage to other plants.

White ants are only too numerous. Dragonflies are plentiful, and it has always

**Black-bellied Tern >**
*Sterna acuticauda*

A small river tern with a deeply forked tail and orange bill. Ashy grey above, black below in summer plumage. Forehead, crown, and crest glossy black. Cheeks, chin, and throat white. Seen on large rivers and jheels in large or small flocks quartering the river a few metres above the surface. Call: a clear piping *peuo*. Feeds by plunge-diving and rests in packed flocks on sandbanks. Food is mainly fish. Resident throughout much of the subcontinent east of the Indus, absent in Sri Lanka.

**Black-bellied Tern** *Sterna acuticauda* J.E. Gray
*Birds of Asia*, Vol. IV, Parts XIX–XXIV, by John Gould, 1867–72. Painted by John Gould & Henry C. Richter.

Courtesy Grindwell Norton Ltd.

Desert Wheatear *Oenanthe deserti* (Temminck)
*Birds of Asia*, Vol. III, Parts XIII–XVIII, by John Gould, 1861–66. Painted by John Gould & Henry C. Richter.

Sponsored by The Raja Bahadur Motilal Poona Mills Ltd.

been a puzzle to me where they can all come from in such a dry locality. They breed, of course, in water, and, though there is only one small stream here, and a few wells, yet at whatever time of the year I go into my garden, I can always see several species. Ant-lions are plentiful, and their pitfalls may be seen almost anywhere. I notice that the Cambridge Natural History states that "The imago is considered to be carnivorous." This I can confirm, as I have frequently seen a species of *Myrmeleon* (Ant-lion) common at Deesa catching small moths and beetles round a lamp at night.

Spiders do not seem to be very abundant, especially the larger kinds. The little red velvety species which appears at the beginning of the rains is one of the most striking, I have been told that a decoction of these is used by natives in Kathiawar, and possibly elsewhere, as an aphrodisiac.

In conclusion I may say that I have been able, thanks to a taste for Natural History to pass many a "Long, long, Indian day" without boredom, even in the hot weather. I hope, later on, when I have had an opportunity of comparing my *Hymenoptera* with those in the British Museum and other collections, to supplement this somewhat discursive paper by a more scientific one, in which the new species collected will be described.

From *JBNHS*, Vol. XIII.

### < Desert Wheatear
*Oenanthe deserti*

A typical wheatear perching on the ground or on low bushes in arid open country; sandy in colour with dark wings, and black throat-patch in male; white patch at the base of the tail; flies low and fast over the ground when disturbed. Breeds in Pakistan (Baluchistan) and in Kashmir, Baltistan, Ladakh, Lahaul, Spiti, and NW Nepal between 3,000 and 5,100 m. In winter it migrates to the wide arid plains of NW India. It is a common sight in Pakistan's NWFP, Punjab, and Sind and reaches the latitude of Mumbai down to central Maharashtra and northern Andhra Pradesh.

NOTE: The article describes the aftermath of the great famine of 1899–1900 which played havoc in Western India. In Gujarat the famine is still remembered as Chhappanio Dukal.

From 1821 till the end of the First World War, Deesa was a military base for the British troops. It was selected because of the perennial stream of fresh water, River Banas, in the immediate neighbourhood and also as it was at the junction of the routes to Sind and Rajasthan, Delhi, and Ahmedabad. Its proximity to Mt. Abu was an added advantage. Plenty of big game in the immediate neighbourhood and abundance of migratory and other birds gave plenty of opportunities to hunters and also to naturalists, who arrived with the army. Capt. E.A. Butler who was stationed here wrote a series of articles in *Stray Feathers* on the avifauna of Mount Abu and North Gujarat and gave an extensive list of the migratory and local birds seen in the area. Lieut. H. Edwin Barnes, a keen birdwatcher and an early member of the Society, was also stationed here. Barnes wrote the popular *Handbook of the Birds of Bombay Presidency* in 1883. Another prominent ornithologist of the Victorian era, A.O. Hume has mentioned Deesa as a paradise for bird watchers. As a result of extensive destruction of forests in recent times, most of the wild animals have vanished. The River Banas which flowed from time immemorial dried up three decades ago, as did many other rivers in Gujarat. Ornithologists too have forgotten this place, but the birds have not. During a recent visit, members of the Society were convinced that Deesa and its immediate neighbourhood can still be a favourite spot for birdwatchers. – A.S.K. & B.F.C.

COURTESY ARUNA & CHANDRAKANT AMBALAL PATEL,
HOLIDAY INN EXPRESS HOTEL & SUITES, HOUSTON, TEXAS, USA,
IN MEMORY OF RIVER BANAS

# Vanishing Wildlife of Panchmahals

James Campbell

As late as the seventeenth century (1616 and 1645) the Dohad [Dahod] forests were famous for their wild elephants. (In 1616 the emperor Jahangir came to Gujarat to hunt elephants in the Dohad forests, and in 1645, seventy-three elephants were caught in the Dohad and Champaner forests.) And twenty years ago, though all traces of wild elephants had passed away, the Panch Mahal and Rewa Kantha districts were, besides of deer and other smaller animals, a favourite resort of tigers, panthers, and bears. Found to some extent over the whole district the larger sorts of game were commonest in Godhra, in parts of Halol, and along the western borders of Dohad and Jhalod. Their favourite haunts were river-bed patches of bastard cypress, Tamarisk, and especially near Godhra the caves and crevices of the low boulder-covered granite hills. The tillage area was then small, and besides stray cattle a fair stock of *nilgai*, small deer, and pig, and a chance spotted deer or stag furnished plentiful supplies for the large beasts of prey. Their quiet was little disturbed, European

**"Rival Monarchs"**. Drawn by Robert Armitage Sterndale. *Denizens of the Jungles*, 1886.

sportsmen seldom visited the district, and from the Bhils and Kolis, except on the rare occasion of some big hunting party, the larger animals had little to fear. In 1860, when the district came under British management, the forests were full of big game, and during the next eight seasons from forty to seventy head were yearly killed. In 1865 the results of the year's shooting included twenty-two tigers, ten panthers, and thirty-eight bears. Besides this destruction, two causes, the clearing of their former haunts, and the shortening of their food supplies, have been at work to reduce the number of big game. Tillage has steadily spread, and not only the open glades, but many thick rich patches of wood on the banks of streams, where tigers used always to lie, are now well guarded fields of tobacco and sugarcane. At the same time greater care in grazing cattle and the destruction of deer have cut down two of the chief sources of their food supply. Tigers are gradually withdrawing from their old haunts. Even in the thickest and safest covers a stray animal is only occasionally found. Panthers wanting less food and shelter give ground slower. But on them too the spread of tillage presses hard, and their numbers steadily drop off. The Pavagad forests and the well-wooded country between Pavagad and Devgad-Bariya still attract the largest game. But even when found, animals take shelter in caves and rocky fissures so deep that neither smoke nor fireworks can drive them out. A sportsman willing to work will probably not leave altogether empty-handed. But blank days will be the rule and success the exception. During the last four years (1874–1877) not more than ten head of large game have on an average been killed. Of Tigers, *vagh*, *Panthera tigris*, two were killed in 1873, six in 1874, and three each in 1876 and 1877. The Panther, *dipdo*, *Panthera pardus*, is still in considerable numbers. But the shelter among the large granite rocks is so good, that once among them panthers are very hard to dislodge. Two were shot in 1876 and four in 1877. The Hunting Leopard, *chita*, *Acinonyx jubatus*, less common than the panther, is sometimes seen. The Black Bear (Sloth Bear), *rinchh*, *Melursus ursinus*, is found in considerable numbers. Like panthers, bear find such good shelter among the granite rocks that they are not often killed. The Hyaena, *taras*, *Hyaena hyaena*; the Jackal, *sial*, *Canis aureus*; and the Fox, *lokri*, *Vulpes bengalensis*, are common everywhere; the Caracal, *siagosh*, *Caracal caracal*, and the Wild Cat, *bangad belli*, *Felis chaus*, are comparatively rare. The Wild Boar, *dukar*, *Sus scrofa*, is found everywhere in the forests. Of deer there are the *sambar*, *Cervus unicolor*, found only on the slopes of Pavagad hill; Spotted Deer, *chital*, *Axis axis*, common in certain parts of the district; the Four-horned Antelope, *Tetracerus quadricornis*, found in most places, and the Gazelle, *chinkara*, *Gazella bennettii*, and Blue Bull, *nilgai*, *Boselaphus tragocamelus*, common everywhere. The Antelope or Blackbuck, *kaliar*, *Antilope cervicapra*, common over the rest of Gujarat is, perhaps because the country is not open enough, scarcely ever found in the Panch Mahals. (Two, the first on record, were shot in 1878. – Mr. Acworth.)

The number of deaths reported from snake-bites was forty-three in 1872, forty-four in 1875, thirty-nine in 1876, and sixty-four in 1877. In Gujarat, Government rewards are granted for the destruction of the following animals: Tigers, full grown, £2 8*s*. (Rs. 24); half grown, £1 4*s*. (Rs. 12); cubs, 12*s*. (Rs. 6); Leopards, Panthers, and Chitas, full grown, £1 4*s*. (Rs. 12); half grown, 12*s*. (Rs. 6); cubs, 6*s*. (Rs. 3); Cobra de capello, 6*d*. (4 annas); *Phursa* or Cobra Manilla, 3*d*. (2 annas); other species possessing a fang in the upper jaw, ¾*d*. (6 pies). The animals are identified and the rewards generally paid by the Mamlatdar.

From *Gazetteer of the Bombay Presidency*,
by James Campbell, Vol. III, *Kaira & Panchmahals*, 1879.

COURTESY GARDEN SILK MILLS LIMITED

# Trees and Plants of Rewa Kantha

James Campbell

A great part of the Rewa Kantha is forest land. The chief trees are the *mahuda*, *Madhuca indica*, found in the greatest plenty in the districts of Chhota Udepur and Bariya. The timber is much used in house building; the flowers are a chief article of food and drink for the poorer Bariya and Udepur tribes, and from the seeds or berries called *doli*, the *doliu* oil is extracted. Teak, *sagvan* or *sag*, *Tectona grandis*, is abundant, but except in *malvans* or sacred village groves is stunted. The timber is used for house building, the seeds and flowers are given in cases of colic, and the leaves are made into thatch. Blackwood, *sisam* or *sisu*, *Dalbergia latifolia*, is not found in any large quantity. Tamarind, *amli*, *Tamarindus indica*, is plentiful, the timber used for house building, and the fruit for pickling. The Mango, *amba*, *Mangifera indica*, is chiefly valued as a fruit tree. Of the Bamboo, *vans*, *Bambusa bambos*, the poles are used for roofing, the young shoots are pickled, and the wheat-like seed is ground into flour and made into bread. The *Rayan*, *Mimusops hexandra*, is abundant and valuable. Its tough wood is used in making native sugar mills and mortars, and in the hot season large number of the poorer tribes feed on its fruit. *Sadado*, *Terminalia arjuna*, timber is largely used in house building and for other purposes. Of the *Khakharo*, *Butea monosperma*, the leaves are made into platters, the flowers called *kesuda* are used as a dye, and the wood for fuel. Its gum serves the place of Indian *kino*. It is given in cases of chronic diarrhoea and is an external astringent application. Of *Beheda*, *Terminalia belerica*, the fruit used as a dye is astringent and forms an ingredient in the compound powder used by native doctors, and called *triphala*. Of the *Timburrun*, *Diospyros embryopteris*, the fruit commonly eaten is believed to lessen the effects of opium. Its wood is hard and is the *abnus* or ebony employed in making boxes and other articles of household furniture. *Bili*, *Ægle marmelos*, is sacred to Shiv, over whose image its leaves are strewn. Its fruit when dry is made into snuff boxes. The pulp of the unripe fruit is useful in cases of dysentery and chronic diarrhoea. *Charoli*, *Buchanania cochinchinensis*, seeds are a favourite native spice. *Dhavdo*, *Anogeissus latifolia*, wood is used for fuel and the gum is mixed with some medicinal drugs and eaten as a cold weather tonic. *Gugali*, *Boswellia serrata*, a sweet-scented gum, is burnt in religious ceremonies, and sometimes used to strengthen lime. *Alardi*, *Morinda exserta*, wood is used for fuel, and the leaves are given to cattle when grass and forage are scarce. *Kher*, *Acacia catechu*, timber is valuable not suffering from water, useful as fuel, and yielding the astringent substance called *kath*, *Terra japonica*. In Bariya, during February and the three following months, *kath* making gives employment to a large number of Kolis and Naikdas. Branches stripped of their bark are cut into small three or four inch pieces and boiled in earthern pots till only a thick sticky decoction remains. A narrow pit five or six feet deep is dug and a basketful of the extract placed over the pit's mouth, the water soaks into the earth and the refuse remains in the basket, leaving the *kath* in the pit. The extract is then taken out of the

**Mangrove Pitta >**
*Pitta megarhyncha*

A brightly coloured dumpy, stub-tailed terrestrial bird. Resembles Indian Pitta with black stripe through eye and white throat, buff breast and flanks. Below ruddy buff with broad bright crimson stripe down abdomen and vent. Differs from Indian Pitta in being much larger in size with larger, longer bill. Crown is uniform rufous-brown. Upper back dull green while lower back, parts of wing, and tail ultramarine blue. Sexes alike. Frequents thin tree jungle with sparse undergrowth, mangrove swamps, gardens, etc. Resident and migratory in Sunderbans, Bangladesh, Myanmar, Malaysia.

Mangrove Pitta (Blue-winged Pitta) *Pitta megarhyncha* Schlegel
*Birds of Asia*, Vol. I, Parts I–VI, by John Gould, 1850–54. Painted by John Gould & Henry C. Richter.

Courtesy Mr. Kavinbhai Parikh, Parikh Foundation

pit and dried on leaves in the sun. The *kher* also yields a white powder called *khersal* given to cure coughs. The soft wood of the *kaledi* tree is made into wooden plates and used for fuel. *Kalam* or *kadam*, *Mitragyna parvifolia*, sacred to Krishna, is used for house building. *Haldharvo*, *Haldina cordifolia*, soft and yellowish is also a useful timber. The Nim, *limbdo*, *Azadirachta indica*, is sawn into planks and used for house building. Its bark serves for cinchona and the leaves are used in fomenting swollen glands, bruises, and sprains. The expressed oil of its seeds is used in cases of leprosy. *Piplo*, *Ficus religiosa*, and *Vad* or Banyan tree, *Ficus bengalensis*, are common. Of the Wood Apple tree, *kothi*, *Feronia limonia*, the fruit is eaten ripe or pickled, and the astringent pulp is given in cases of diarrhoea and dysentery. *Moheno* wood is used for fuel. *Tanach*, *Ougenia oojeinensis*, wood is tough and used in cart building. *Baval*, *Acacia arabica*, wood is used for fuel and in making cart wheels. Its gum is valuable and its astringent bark is used in tanning. Of the Palmyra, *tad*, *Borassus flabelliformis*, the juice yields toddy and the leaves serve for thatching. The juice of the Wild Date, *khajuri*, *Phoenix sylvestris*, yields toddy, and its fruit is eaten by the poor classes. Blunt-leaved Zizyphus, *bordi*, *Zizyphus mauritiana*, fruit is eaten, and is a favourite food with bears. *Samdi*, *Prosopis spicigera*, is worshipped on the Dasera festival (September-October). Its pods called *sangri* are used as vegetables. Custard Apple, *sitaphal*, *Anona squamosa*, is chiefly valued for its fruit. *Kanji* or *karanj*, *Pongramia pinnata*, yields an oil useful in cases of itch and burning. *Rohen*, *Soymida febrifuga*, bark yields a dark red dye. It tastes bitter and may be used like Peruvian bark. A good tonic in intermittent fever, it causes dizziness if too much is taken. *Kudo*, *Wrightia tinctoria*, flowers are mixed with curry and taken as a vegetable. The seeds called *indrajav* are useful in dysentery. The bark, formerly exported to Europe under the name of Concan or Tellicherry bark, is astringent and bitter and is employed in fever and dysentery with much success. [Editors' note: Concan or Tellicherry bark comes from *Holarrhena antidysenterica*, not *Wrightia tinctoria*.] *Sevan*, *Gmelina arborea*, a light wood, is used in making carts and some articles of furniture. *Simlo*, *Bombax ceiba*, wood is soft and is hollowed into canoes or small boats. The fine cotton-like wool that covers its seeds is used for stuffing pillows, and its gum, called *kamarkas*, ground to powder is drunk in milk as a tonic. *Pilu*, *Salvadora persica*, berries are aromatic and pungent to the

"The Sacred Hindoo Grove near Chandod on the Banks of the Nerbudda". Drawn by James Forbes, 1782. *Oriental Memoirs*, Vol. III, 1813.

**Blue Pitta >**
*Pitta cyanea*

A large pitta with forehead and crown greenish grey changing to scarlet on hind crown, black stripe through eye, black moustachial streaks on either side of whitish throat, and black spotting and barring on under-parts. In flight shows small white patch in wing feathers. Adult male has blue upper-parts and tail, and pale blue wash to under-parts. Adult female has dark olive upper parts with blue tinge. Affects damp ravines and scrubby undergrowth in mixed tree and bamboo forest in evergreen forest. Keeps singly on the ground in undergrowth, hopping about, turning over or flicking aside the dry leaves, and digging into the damp soil with its bill for food. When alarmed disappears into cover by long swift jumps like small rodent, or flies up into a tree whence it soon descends again. Resident, NE India, Bhutan, Bangladesh, Myanmar, Thailand, etc.

Blue Pitta *Pitta cyanea* Blyth
*Birds of Asia*, Vol. I, Parts I–VI, by John Gould, 1850–54. Painted by John Gould & Henry C. Richter.

Courtesy Premchand Ishwarlal Shah Family, Venus Jewel

taste. *Rohido, Tecomella undulata,* is supposed to cure a swelling in the belly, and the disease known among native doctors as congealed blood. A tree of this kind is kept with great care by the Raja of Rajpipla. *Agathio, Sesbania grandiflora,* flowers are used for food and the bark as a tonic. The seeds of the *Arithi, Sapindus trifoliatus,* known as soapnuts, are used in cleaning the hair.

The following are some of the principal shrubs and medicinal plants found in the Rewa Kantha forests. *Agheda, Achyranthes aspera,* the seeds are given in cases of hydrophobia and snake-bite, the juice of its flowering spike for scorpion bites, and the ashes of the burnt plant have been successfully used in dropsy. *Gorakh amli, Adansonia digitata,* the pulp is a good refrigerant in fever, and the bark a useful substitute for quinine in low fever. *Kariaturi, Andrographis paniculata,* an infusion of its leaves is used as a tonic and febrifuge. *Samudra shok, Argyreia speciosa (A. nervosa),* the leaves are used to foment boils and abscesses. *Shatarasi (Shatavari), Asparagus racemosus,* the root when fresh is a mild tonic. *Gokhru, Tribulus terrestris,* the root is a tonic and diuretic. *Dholi satardi, Boerhaavia diffusa,* the root is said to be a strong emetic. *Eranda kakdi* (papaya), *Carica papaya,* the milky juice is reckoned one of the best vermifuges. *Garmala, Cassia fistula,* the pod pulp acts as a strong purgative. *Indrak, Citrullus colocynthis,* the pulp of the fruit is purgative. *Dholo akdo, Calotropis gigantea,* the root-bark is used as a diaphoretic, an emetic in large doses, and as an alterative in leprosy. [Editors' Note: The species found in Gujarat is *C. procera.*] *Musli (Kali musli),*

**"The Wedded Banyan Tree, or the Palmyra and the Burr Tree United"**. Drawn on Salsette by James Forbes, 1774. *Oriental Memoirs,* Vol. II, 1813.

*Curculigo orchioides*, the root slightly bitter and aromatic is used in gonorrhoea. *Amarvel, Cuscuta reflexa*, the stem is used as an alternative, especially in bilious disorders. *Nagar moth, Cyperus rotundus*, the fresh tubers are a stimulant and diaphoretic. *Jangli suran, Dracontium polyphyllum*, the roots are used as an antispasmodic in asthma. *Kalu ganthi (Bhangro), Eclipta prostrata*, the root is a purgative and emetic, used in cases of enlarged spleen, liver, and dropsy. *Thor, Euphorbia nerifolia*, the milky juice is given as a purgative, and is put in the ears to cure ear-ache. *Pitpapdo, Fumaria parviflora*, the whole plant is used with black pepper in common agues. It is said to be a diuretic, diaphoretic, and aperient. *Sagargot, Caesalpinea crista*, the kernels of the nut are very bitter and powerfully tonic. They are given in the form of powder mixed with spices in intermittent fever. *Brahmi, Hydrocotyle asiatica*, the whole plant is considered diuretic. It is a good alterative and has been used with success in skin diseases. It is said to cure brain disorders. *Aduso, Adhatoda zeylanica*, the juice of the leaves and flowers, expectorant and antispasmodic, is given in chronic bronchitis and asthma. *Bhui champo, Kaempferia rotunda*, the roots are stomachic and applied to swellings. *Tumbdi, Leucas linifolia*, in snake-bites the leaves are bruised and a teaspoonful of the juice given to be inhaled through the nostrils. *Bhui amli, Phyllanthus erecta*, the roots, fresh leaves, and young shoots are used as diuretic, the roots and fresh leaves in jaundice or bilious complaints, and the young shoots as an infusion in dysentery. *Isapgol, Plantago ispaghula*, the seeds mucilaginous and demulcent may, mixed with sugarcandy, be given in the form of a cold infusion thrice a day in cases of dysentery and gonorrhoea. *Lal chitrak, Plumbago zeylanica*, the fresh bark is made into a paste and applied to indolent buboes and tumours. *Bavchi, Cullen corylifolia*, the seeds aromatic and slightly bitter are said to be stomachic and are used in cases of leprosy and other skin diseases. *Gajkarni, Rhinacanthus nasuta*, the juice of the leaves and roots is applied as a cure for ringworm. *Mundavli, Sphaeranthus indicus*, the seeds considered to cure worms are prescribed in powders. The powdered root is stomachic, and the bark powdered and mixed with whey is a valuable remedy for piles. *Gulvel, Tinospora glabra*, the stem is a good tonic and diuretic. A cold infusion has been found to be of much benefit in chronic rheumatism and remittent fever. *Kalijiri, Baccaroides anthelmintica*, the seeds are very bitter and powerfully anthelmintic and diuretic. Reduced to powder and mixed with lime juice they are used to destroy lice. *Nagod, Vitex nigundo*, the roots are used as a decoction, as a vermifuge, and as a diaphoretic in protracted fevers. *Dhavdi, Woodfordia fruticosa*, the flowers are powerfully astringent. A decoction is used in cases of diarrhoea. *Malkangni, Celastrus paniculata*, the oil of the seeds is a diuretic and has been used successfully in healing sinuses and fistulae. *Hansraj, Adiantum philippense*, the leaf of this fern is used in cases of fever and cough. *Balbaja* is used for ascites occurring in children. *Kolijan, Alpinia galanga*, the root is used in cases of cough and rheumatism. *Gani*, the seed is used in constipation. *Nilophal, Nymphoea nouchali*, also called *poyana*, is used generally in the form of a syrup in cases of fever. *Ram tulsi, Ocimum gratissimum*, is used for headache, fever, pain of the intestines, and colds. *Nirgundi, Vitex negundo*, the fruit is used for gleet and debility. *Cheran*, considered a good tonic, is said to heal broken bones. Among Rewa Kantha grasses the most important are *viran* or *khas, Vetiveria zizanoides*, when wetted a well known screen for cooling hot winds, and elephant grass, *baru, Coix lachryma-jobi* whose stems are used for native pens, *kalam*.

From *Gazetteer of the Bombay Presidency*,
by James Campbell, Vol. VI,
*Rewa Kantha, Narukot, Cambay, & Surat States*, 1880.

COURTESY DIPIKA & SHASHIKANT DOSHI, HONEY & ASHISH DOSHI,
CHETNA & SHRUTI, OOPAL DIAMOND

# Khandesh, a Stronghold of Wild Beasts

JAMES CAMPBELL

Up to the seventeenth century, the hilly tracts to the north of Khandesh were a great breeding place for wild elephants. (In 1630 Jamalkhan Karawal came to the Gujarat-Khandesh frontier and captured 130 elephants in the Sultanpur forests, of which 70 were sent to Delhi. – *Watson's Gujarat*, p. 71.) But probably from the frequent passage of armed bodies during the Moghal conquest of the Deccan, from the increase of traffic down the Tapti valley to Surat, and from the spread of tillage in Khandesh they were, during the eighteenth century, frightened off. The chief wild animal still found in the district is the Tiger, *vagh*, *Panthera tigris*. In the disturbed times at the beginning of the present century, large tracts passed from tillage into forest, and tigers roamed and destroyed in the very heart of the district. In 1822 wild beasts killed 500 human beings and 20,000 head of cattle. Their destruction was one of the most pressing necessities, and in May, June and July of that year (1822), as many as sixty tigers were killed. In spite of the efforts of Sir James Outram and his successors, tigers and other large beasts of prey continued so numerous that the fear of them kept waste and desolate some of the richest tracts in Khandesh. Even as late as the mutinies (1857–1859), Khandesh, more than almost any part of western India, continued a stronghold for wild beasts. So dangerous and destructive were they that a special division of the Bhil corps were, as tiger hunters, set apart to aid the Superintendents of police. Since 1862, under the Superintendent of Police Major O. Probyn, the destruction of tigers has gone on apace. Of late years, to the efforts of the district officers have been added a rapid spread of tillage and increase of population. The tiger is no longer found in the plains. Among the Satpudas in the north, along the Nemad frontier and the Hatti hills in the east and the south-east, in the Satmalas in the south, and in the Dangs and other wild western tracts he still roams. Even there his number is declining. The loss of cattle is inconsiderable and the loss of human life trifling. In the five years ending 1879, sixteen human beings and 391 head of cattle were killed by them. The returns show a fall in the number of tigers slain from an average of nearly fifteen in the five years ending 1870 to ten in the nine years ending 1879.

The Panther, *bibla* or *bimta*, *Panthera pardus*, is generally said to be of three distinct species, two large and one small. Of the two large kinds, one rivals the tigress in size, and as he will attack unprovoked, is equally or even more dangerous to man; the other smaller, stouter, and with a round bull-dog's head, has a looser, darker, and longer fur, with spots much more crowded and quite black along the ridge of the back and up the legs about as high as the shoulders and thighs. The third variety is a very different animal, much smaller and darker. As it lives chiefly on dogs, it is known among the natives as the dogslayer, *kuttemar*. In the fifteen years ending 1879, 658 panthers were killed, the yearly number varying from seventy-eight in 1878 to nineteen

---

**Bright-headed Cisticola >**
***Cisticola exilis tytleri***
A diminutive warbler of grassy hill slopes. Male (summer): above, crown unstreaked pale orange-yellow; back rufous-brown black-streaked. Tail black, tipped with buff. Below, centre of belly white; rest ochre-buff. Female and winter male: above, crown and back rufous streaked with black. Tail brown above with black band and buff tip, greyish below. This warbler is found in suitable plateaus of grassland from foothills of 400 m up to 1,500 m. It collects in loose parties of ten or twenty pairs. They remain hidden in grass cover, individuals mounting a grass stem from time to time with tail constantly flicked open like a fan. When disturbed the bird jerks itself straight into the air and after flying for some distance falls headlong into the grass again. Stuart Baker writes, "It has a call-note sounding like *chir-r-r-r*, and after an interval, a beautiful bell like tinkle which seems to come from a different direction. Its food consists of ants and minute insects found among the grass stems." Resident, N. Uttar Pradesh, Nepal terai east to Bhutan duars, Meghalaya, Assam, Arunachal Pradesh, NE Manipur, Nagaland, and Bangladesh. In winter descends to the plains of Assam and W. Bengal.

**Bright-headed Cisticola (Yellow-headed Fan-tailed Warbler)**
*Cisticola exilis tytleri* Jerdon

*Fauna of British India*, Vol. II, by W.T. Blanford, 1924. Painted by E.C. Stuart Baker.

In Memory of Rutty C. Dady. Courtesy Dady C. Dady & Diana Ratnagar

in 1870. The Hunting Leopard, *chita, Acinonyx jubatus*, quite a different animal from the panther, has, like a dog, claws that do not draw in. In form like a greyhound, it has a short mane, bushy black-spotted fur, and a black tail. It is very rare in Khandesh, found in the Satpuda hills only. The Wild Cat, *ran manjar, Felis chaus*, met all over the district, is comparatively harmless, and differs in size, colour, and length of tail, only slightly from the house cat. The Lynx, *Caracal caracal*, a rare animal, is occasionally found among rocky hills. It is very shy, and is seldom abroad after daybreak.

The Hyena, *taras, Hyaena hyaena*, once very common, is now rarely seen. The Wolf, *landga, Canis lupus*, formerly caused much havoc among sheep and goats, and is even known to have carried off young children. Like the other flesh-eaters, he has been forced to give way before the spread of tillage. Still he is very destructive, and though he seldom attacks human beings, kills an immense number of sheep and goats, and two or three together will often pull down a good-sized young buffalo or heifer. During the fourteen years ending 1879, 4,138 wolves were killed, the yearly number varying from 603 in 1874 to seventy-one in 1879. Besides the above, the Jackal, *kolha, Canis aureus*, and the Fox, *khokad, Vulpes bengalensis*, abound in the open country. The Wild Dog, *kolsunda, Cuon alpinus*, is also found in the Satpuda hills, hunting in packs.

The Sloth Bear, *asval, Melursus ursinus*, is found in all the forest-clad hills of Khandesh. Formerly abounding in the rocky hill tops of Pimpalner and Baglan in the south-west, the number of black bears has during the past twenty years been much reduced. Though not generally dangerous to life, he is at times very mischievous. Sugarcane, when he can get it, is one of his favourite articles of food, and he destroys much more than he eats. The flower of the *moha, Madhuca indica*, tree is his chief sustenance at the beginning of the hot season. This flower, which produces the

"Meaning Mischief". Drawn by Robert Armitage Sterndale. *Denizens of the Jungles*, 1886.

common spirit of the country, seems to affect the bear with a kind of intoxication, as he is known to be most dangerous at that season, and apt to attack man unprovoked. A vegetarian, except as regards ants and some other insects, he does no injury to flocks or herds.

The Hog, *duckker*, *Sus scrofa*, of all wild animals, causes most loss to the cultivator. Though, save in the set of his tail, much like the domestic village pig, he differs from him widely in habits. A pure vegetable eater, he is most dainty in his tastes. He must have the very best the land affords, and while choosing the daintiest morsels, destroys much more than he eats. Sugarcane, sweet potato and other roots, and juicy millet and Indian corn stalks are his favourite food. A few years ago herds of wild pig were found everywhere, but their numbers are now much smaller. From the border hills they still sally at night to ravage the crops in the neighbourhood, but they are no longer so destructive as they once were. With the aid of their dogs and spears, the Bhils hunt and kill them for food, and the clearing of the forests has made their destruction comparatively easy. Twenty years ago in the country east of the Purna river, then belonging to His Highness Sindia (of Gwalior), herds of some hundreds might be seen marauding in open day. Night and day the cultivator had to watch his

**"The Jack Tree (Jackfruit Tree), the Man and Fruit in Proportion"**. Drawn by James Forbes, Bombay, 1767. *Oriental Memoirs*, Vol. I, 1812. The jackfruit tree (*Artocarpus heterophyllus*) is a large evergreen tree, with a dense crown of dark green leaves. It bears the largest edible fruit in the world, sometimes 45 kg in weight. The jackfruit is native to the forests of the Western Ghats and is cultivated in the warmer regions of India, Myanmar, and Sri Lanka.

fields. Though comparatively few are left, herds of fifty and upwards are still occasionally seen.

The Bison, *gava, Bos gaurus*, is found only in the Satpuda and Hatti hills. The shyest and wariest of forest animals, its chief food is grass and young bamboo shoots. The Stag, *sambar, Cervus unicolor*, is found in all the hill country on the borders of the district. It feeds in the plains and fields at night, and seeks the hill tops at early dawn. It seldom, if ever, lies in the plain country. The Spotted Deer, *chital, Axis axis*, is now rare. He is never found far from winter, and generally in thick forests. In the country east of the Purna spotted deer were formerly found in immense numbers, but most of them were shot or driven away while the railway was making. They are still in small numbers near rivers in the Satpuda hills, and in the western forests along the Tapti. The Barking Deer, *bhekre, Muntiacus muntjak*, and the Four-horned Antelope, also called *bhekre, Tetracerus quadricornis*, are occasionally met with in the Satpuda hills. The Blue Bull, *nilgay, Boselaphus tragocamelus*, was once common everywhere, but is now confined to the few strips of forest land left between the Satpuda and other hills and the open plains, and to the low country on the west. He seldom enters the hills or dense forests, feeding chiefly on *palas, Butea monosperma*, or other trees in the flat country. The Indian Antelope or Blackbuck, *kalvit, Antilope cervicapra*, frequents the open fields and devours the corn. Disliking forest country, they were never so plentiful in Khandesh as in the Deccan and Gujarat plains. Very few of them are left. The Indian Gazelle, *chinkara, Gazella bennettii*, loving the shrub brushwood and rocky eminences of Khandesh, are still comparatively plentiful. The Common Hare, *sasa, Lepus nigricollis*, found in considerable numbers all over the district, completes the list of four-footed game animals.

From *Gazetteer of the Bombay Presidency*,
by James Campbell, Vol. XII, *Khandesh*, 1880.

COURTESY YASHWANT DATTATREYA JOSHI, ANDHERI (E), MUMBAI

# Wildlife of Jalpaiguri District

John F. Gruning

The Jalpaiguri district has always been famous for its big game and, though the heavy grass and reed jungle which is the favourite resort of wild animals is steadily diminishing owing to the extension of cultivation, the sanctuary afforded by the numerous reserved forests will prevent game from being killed out and the district will always afford good sport.

Among the larger carnivora are the tiger, the leopard and the clouded leopard. The tiger is found all over the Western Duars, in the neighbourhood of the forests; the most famous shooting-ground is on the east bank of the Jaldhaka river opposite Ramshai Hat where Lord Curzon shot several tigers in 1904. Tigers are also occasionally seen west of the Tista and one was shot in 1907 in a small patch of scrub jungle about four miles from Jalpaiguri, not far from the southern extremity of the Baikanthpur forest. Man-eaters are almost unknown; in the few cases in which human beings have been killed by tigers, the corpses were left untouched; game and cattle are so numerous in the district that tigers are not driven to eat human flesh. The largest tiger, which has been shot in Western Duars, measured 10' 2". Leopards are common all over the district, any small patch of scrub jungle gives them cover and they do much harm to the villagers by carrying off their cows, goats, pigs and dogs; they are far bolder than tigers and attack with less provocation. On one occasion the Assistant Manager of a tea-garden was riding a bicycle along a well frequented road, when, from a patch of jungle close to the tea, a leopard sprang on him, knocked him off his machine, and clawed him badly. On another, a Mech, cutting firewood in the jungle was attacked by a leopard, which he killed with his dao (long knife) after a hard struggle; he was brought into the hospital at Alipur Duar very badly mauled about the head but recovered after some months. The clouded leopard is very rare and is found only in the Buxa hills. A black leopard was shot about five miles from Jalpaiguri in 1906 by the Superintendent of Police. The leopard cat and the jungle cat are common, as are also the larger civet cat and the smaller civet cat. The genus *Canis* is represented by the jackal and the genus *Cuon* by the wild dog. Wild dogs are seldom met with and no report of damage done by them has been received in recent years. The only representative of the genus *Vulpes* is the Indian fox.

The order Ungulata comprises the elephant, the wild pig and various Ruminantia including the rhinoceros, bison, wild buffalo, and many kinds of deer. Elephants are found in considerable numbers throughout the forests and appear to have increased in recent years; they come down from the hills in large numbers about the time when the rains break in June and again in November when the rice crops are ripening, on which occasions they do considerable damage. Solitary males, both tuskers and muknas, are a serious menace to life in the tracts through which they roam, and no less than five have been proclaimed in the last two years. One of these, a tusker, appeared at Madari Hat in March 1905; he pulled down several houses, charged the engine-shed, making a large hole in the masonry wall, damaged a first-class carriage standing in the railway station, and injured several people. He was next heard of at the Hantapara Tea-garden where he killed a woman, after which be disappeared and was at last shot in December 1907 by the Assistant Manager of the Chuapara Tea-garden, where he had chased the labourers from their work. On the road through the forest to Buxa, it was found impossible to use telegraph posts as the elephants pulled them down as fast as they were put up, and the wire is now attached to large trees. The Manager of the Bengal-Duars Railway also complained of telegraph posts along the line through the forest between Latiguri and Ramshai Hat stations having been pulled down by elephants. The wild pig is common throughout the district and its flesh is eaten by Rajbansis, Meches, Garos and Nepalis. Rhinoceros, buffalo and bison were in danger of being shot out, and, to prevent their extinction, they are now protected in the reserved forests. The *Rhinoceros unicornis*, *Rhinoceros sondaicus* and *Didermocerus sumatrensis*, are all found in the district; the last named is very rare but has been shot in the Dalgaon forest. The *Rhinoceros unicornis* appears to be increasing and I have myself seen over twenty fresh rhinoceros beds while shooting in a grass jungle north of Silitorsa. Buffalo are not numerous but bison (*Bos gaurus* and *Bos frontalis*) are found from time to time. Of the deer tribe, the sambhar is often seen in the forest, hog deer, swamp deer and barking deer are still common in the district though their numbers are decreasing as cultivation extends. A few spotted deer or chital are still to be found in the forests to the north of the Alipur and Bhalka tahsils.

The Ursidae are represented by the Himalayan black bear and the common Indian sloth bear. The Himalayan black bear is fierce and readily attacks without provocation anyone who gets in its way; it is not uncommon to hear of villagers being killed by this bear and in Mech villages men may often be seen who have been mauled badly. Mr. Ainslie, the Subdivisional Officer of Alipur Duar, who has shot several, told me that he has never seen a tiger fight so hard as one of these did; it charged the beating elephants, seized one of them by the hind leg, and went on charging and fighting till it was killed. Other mammalia found in the district are the common Indian hare, the hispid hare which is very rare, monkeys, squirrels, otters, porcupines and several of the smaller rodents.

**Babur Hunting Rhinoceros >**

In the 16th century rhinos were found as far north as Peshawar and Sind. This painting from the *Babur Nama* describes a hunting scene dated 10th December, 1526 near Bigram (Peshawar). Babur, the founder of the Mughal dynasty, crossed the river Siyah-ab and formed a hunting circle downstream. He records in the *Babur Nama*, "After a little, a person brought word that there was a rhino in a jungle near Bigram, and that people had been stationed near-about it. We betook ourselves, loose rein, to the place, formed a ring around the small jungle, made a noise, and brought the rhino out and it rushed across the plain. Humayun and his friends had never seen a rhino before and were much entertained. It was pursued for two miles; many arrows were shot at it; and it was brought down without much resistance. Two other were also killed. I had often wondered how a rhino and an elephant would behave if brought face to face; this time one came out right in front of some elephants the mahouts were bringing along, it did not face them when the mahouts drove them towards it, but got off in another direction."

**" More than His Match"**. Drawn by Robert Armitage Sterndale. *Denizens of the Jungles*, 1886.

Babur Hunting Rhinoceros near Peshawar
Painting from the *Babur Nama* reproduced with the kind permission of the National Museum, New Delhi.

Courtesy Blue Cross Laboratories Limited

Squirrels, a Peacock and Peahen, Sarus Cranes, and Fishes
Painting from the *Babur Nama* reproduced with the kind permission of the National Museum, New Delhi.

Courtesy Madans & Dhingras

**< Squirrels, etc.**
The *Babur Nama* in the National Museum, New Delhi, has 378 folios. Of these 122 are illustrated and 42 of the illustrations depict flora and fauna. Babur starts his account of the birds of India with the Peacock: "The peacock is a beautifully coloured and splendid bird. Its body may be as large as a crane's but it is not so tall. On the head of both cock and hen are 20 or 30 feathers rising some 2 or 3 inches high. The hen has neither colour nor beauty. The head of the cock has an iridescent collar; its neck is of a beautiful blue; below the neck, its back is painted in yellow, parrot green, blue and violet colours. The flowers on its back are much the smaller; below the back till the tip of the tail are larger flowers painted in the same colours. The tail of some peacocks grows to the length of a man's extended arms. Its flight is feebler than the pheasants; it cannot do more than one or two short flights. Hindustani call the peacock mor." This painting is by Bhawani, who excelled in painting birds and animals. At the top squirrels are playing on a tree. In the middle, a peacock and a peahen are shown, below them a pair of Sarus Cranes, and in the pond a pair of fishes. This is one of the best paintings of nature in the *Babur Nama*.

The following is a list of the larger wild animals found in Jalpaiguri district:–
The tiger (*Panthera tigris*).
The leopard (*Panthera pardus*).
The clouded leopard (*Neofelis nebulosa*).
The leopard cat (*Prionailurus bengalensis*).
The jungle cat (*Felis chaus*).
The larger civet cat (*Viverra zibetha*).
The smaller civet cat (*Viverricula indica*).
The jackal (*Canis aureus*).
The wild dog (*Cuon alpinus*).
The Indian fox (*Vulpes bengalensis*).
The elephant (*Elephas maximus*).
The wild pig (*Sus scropa*).
The rhinoceros (*Rhinoceros unicornis*).
　　(*Rhinoceros sondaicus*).
　　(*Didermocerus sumatrensis*).
The wild buffalo (*Bubalus arnee*).
The bison (*Bos gaurus*).
　　(*Bos frontalis*).
The sambhar (*Cervus unicolor*).
The swamp deer (*Cervus duvaucelii*).
The hog deer (*Axis porcinus*).
The barking deer (*Muntiacus muntjak*).
The spotted deer or chital (*Axis axis*).
The Himalayan black bear (*Ursus thibetanus*).
The common Indian sloth bear (*Melursus ursinus*).
The hispid hare (*Caprolagus hispidus*).
The Bengal monkey or Rhesus macaque (*Macaca mulatta*).
The black squirrel (*Sciurus giganteus* [*Ratufa bicolor?*]).
The grey squirrel (*Sciurus lokriah* [*Callosciurus* sp. or *Dremomys lokriah?*]).
The common Indian squirrel (*Funambulus palmarum*).
The Indian porcupine (*Hystrix indica*).
The otter (*Lutra lutra*).

Game birds used to abound in the Western Duars but many species are getting scarce as the grasslands are being brought under cultivation.

Many varieties of snakes are found in the district while the numerous rivers and streams in the district contain many varieties of fish of which the mahseer, rohu and katli are the biggest.

From *Eastern Bengal and Assam District Gazetteers – Jalpaiguri*
by John F. Gruning, 1911.

# A Visit to Narcondam

B.B. Osmaston, i.f.s.

Narcondam is a small solitary island situated in the Andaman Sea in Lat. 13°-26'. It is well out of sight of land, the nearest being the Great Coco and North Andaman Islands, both about 80 miles distant to the north-west and west, respectively.

The island is about seven miles in circumference and the central peak reaches a height of 2,200 feet above the sea.

It rises abruptly out of a deep sea from over 500 fathoms and its origin is certainly volcanic, though there are no signs of a crater or of any recent volcanic activity.

It having been decided that the forest growth in Narcondam should be explored with a view to ascertain whether any of the valuable Andaman Padauk (*Pterocarpus dalbergioides*) occurred there, I gladly availed myself of the opportunity of visiting such an interesting and unfrequented island. I spent five days, October 1st to 6th, camped on the island in company with my friend C. Gilbert Rogers during which time we thoroughly explored a great part of the island, ascending the central peak, as well as circumnavigating the island in a nine foot canvas canoe.

The whole island is clothed more or less densely with forest from coast line to summit. In places the jungle is almost impenetrable; in others one can move about freely in the dense shade afforded by palms (chiefly *Caryota mitis*) under a lofty canopy of huge forest trees including immense figs but no Padauk or other valuable timber.

Fresh water is not to be found anywhere on the island except at a spot near the north-east corner where there was a small pool in the bed of a stream which however would certainly be dry from November to April. Most of my time was devoted to a study of the birds which, however, I found to be scarce both in species and individuals.

Altogether seventeen kinds were observed, of which at least seven are only seasonal visitors.

The following short account of the birds obtained may be of interest:–

1. *Aceros narcondami* (The Narcondam Hornbill).
This Hornbill is, as is well known, peculiar to Narcondam where it was discovered by Hume in 1873. I found it fairly numerous in the high forest which clothes the lower slopes of the mountain down to the coast.

They are both noisy and fearless and from their conspicuous black and white colouration are bound to attract the attention of the most unobservant.

I found them feeding exclusively on figs, and such trees in fruit formed a centre

**Nicobar Parakeet >**
*Psittacula caniceps*
A large, long-tailed, grass-green parakeet with black head markings. Male (adult), forehead black continued backward as a broad stripe to each eye; a wide black band from lower mandible to each side of neck. Rest of head, hind neck, and sides of neck brownish grey diffusing into the bright yellowish green underparts. Tail dingy yellow; middle feathers blue at base, violet-grey towards tips. Bill red. Female has bluish tinge in grey of head and black bill. Keeps singly, or in pairs, or parties of 5 or 6 much to the tops of high trees in dense forest rendering it difficult to observe. Feeds on ripe fruit of pandanus. Call: a wild screeching note continuously uttered while at rest and in flight: a loud, raucous *kran kran* not unlike a crow's (Humayun Abdulali quoted by Sálim Ali). Resident, endemic to Nicobar Islands. Recorded only on Great Nicobar, Montschall and Kondul (Sálim Ali and S. Dillon Ripley).

Nicobar Parakeet *Psittacula caniceps* (Blyth)
*Birds of Asia*, Vol. II, Parts VII–XII, by John Gould, 1855–60. Painted by John Gould & Henry C. Richter.

Courtesy Tata Consultancy Services Limited

Long-tailed Parakeet (Nicobar Red-cheeked Parakeet) *Psittacula longicauda* (Gould)
*Birds of Asia*, Vol. II, Parts VII–XII, 1855–60. Painted by John Gould & Henry C. Richter.

In Memory of Shri G.S. Kamath & Smt. Suguna G. Kamath,
from Ganesh & Gokul G. Kamath & Family, Gokul Ice-creams, Santacruz (W), Mumbai

of attraction to the birds who resorted to them from far and near. This Hornbill, restricted as it is to an area of under three square miles must be, judging from the number of individuals, one of the rarest, if not the rarest, bird in the world.

At a liberal estimate there cannot be more than 200 Hornbills on Narcondam.

2. *Psittacula eupatria magnirostris* (The Large Andaman Paroquet, Alexandrine Parakeet)
This Paroquet which is so common in the Andamans is also fairly numerous on Narcondam. It keeps chiefly to the top of lofty trees and is difficult to procure.

The only species of Paroquet previously recorded from this island is *P. longicauda tytleri* (the Red-cheeked Andaman Paroquet, Long-tailed Parakeet). I know this species well but neither heard nor saw anything of it.

3. *Nectarinia jugularis andamanica* (The Andaman Sunbird, Olive-backed Sunbird).
This little honey-sucker is the commonest bird on the island. It chiefly frequents the coast.

4. *Ducula bicolor* (The Pied Imperial Pigeon).
This fine conspicuous pigeon is fairly common especially near the shore. Mr. A.O. Hume says it is only a seasonal visitor, though on what grounds I do not know. I should not be surprised to find it proved to be a resident.

5. *Halcyon pileata* (The Black-capped Kingfisher).
I saw only two specimens of this beautiful Kingfisher, so it is presumably rather rare.

6. *Collocalia brevirostris innominata* (Hume's Swiftlet).
I saw a number of these Swiftlets hawking flies around the summit of the mountain. They probably breed in the caves along the south coast of the island.

7. *Egretta sacra* (The Eastern Reef Heron, Pacific/Eastern Reef Egret).
This Reef-Heron is fairly common along the coast.

8. *Haliaeetus leucogaster* (The White-bellied Sea Eagle).
I saw a pair, as well as a young bird in immature plumage.

9. *Astur* (?) species. *Accipiter butleri* (Nicobar Sparrowhawk).
I saw two small hawks circling around the top of the mountain. They resembled *Astur* in their flight.

10. *Chalcophaps indica* (The Bronze-winged Dove, Emerald Dove).
I shot a single specimen of this dove, the only one I saw. It is apparently very rare.

11. *Eudynamys scolopacea dolosa* (The Indian Koel, Asian Koel).
I heard and saw a good many Koel. They are undoubtedly, as in the Andamans, only cold weather visitors.

12. *Hirundo rustica* (The Swallow, Barn Swallow).
Common along the shore and near the summit. Migratory.

13. *Motacilla cinerea* (The Grey Wagtail).
Winter migrant.

---

**< Long-tailed Parakeet**
*Psittacula longicauda*
Adult male has pinkish red cheeks contrasting with dark green crown, broad black chin stripe, nape yellowish green with an indistinct lilac collar on hind neck and variable pale turquoise and lilac wash to green of mantle. Wings largely yellow-green and blue-green. Tail feathers green and blue, middle feathers blue. Under-parts green, yellower on throat and breast. Female has the cheeks faint red and the chin stripe partially tinged green while rest of the plumage nearly uniform green. Food: as other parakeets – grain and fruit. Items specifically recorded are papaya, ripe pandanus, and occasionally the outer covering of betel nuts (Davidson, 1874). Affects forests, gardens, cultivation, and mangroves. Restless; dashes from one tree to another and twists and turns with ease while flying at high speed. Call similar to that of Rose-ringed Parakeet. Abundant on Nicobar group of islands.

14. *Dendronanthus indicus* (The Forest Wagtail).
Winter migrant.

15. *Arenaria interpres* (The Ruddy Turnstone).
Winter migrant.

16. *Actitis hypoleucos* (The Common Sandpiper).
Winter migrant.

17. *Merops philippinus* (The Blue-tailed Bee-eater).
Winter migrant.

Among mammals I found two species of Fruit Bats. The Nicobar Flying Fox (*Pteropus melanotus*) and another smaller species, as well as a rat which appeared to be semi-arboreal in its habits. Of reptiles *Varanus salvator* (The Ceylon Monitor) was very common, especially near the shore. They use their powerful tails in self-defence, inflicting a nasty blow upon any one approaching them incautiously from behind.

Skinks of various sizes were also very common as well as *Calotes* sp. and a beautiful little green lizard provided with suctorial feet.

We also obtained a snake but no frogs or toads.

The ground was, as is usual on such islands, alive with hermit crabs of all sizes, and large whitish land crabs were to be found in holes at the root of some of the larger trees. Mosquitoes were fortunately very rare owing no doubt chiefly to the absence of fresh water.

I saw a couple of scorpions under fallen wood.

The forest belongs to the tropical evergreen type and some of the trees attain very large dimensions both in girth and height. Figs are very numerous, and it is no doubt largely due to this fact that so large a number of fruit-eating birds can support themselves all the year round on so small an area.

Towards the summit of the mountain the tree growth becomes stunted and the vegetation alters markedly in character, such genera as *Strobilanthes*, *Eschynanthus* and *Begonia* being represented, which recall the flora of the eastern temperate Himalayas.

The temperature at the top of the mountain at midday was 74° in the shade, that at the bottom under similar conditions being 82°.

The view from the top was most impressive, commanding as it does the whole of the island, which is spread out at one's feet, surrounded by the limitless ocean.

From *JBNHS*, Vol. XVI.

**Sunda Teal >**
*Anas gibberifrons*

A small mainly brown duck, with white markings on head, white throat, and white narrow eye-ring which helps in identification. Sexes alike. Usually associates in flocks of 20–30. Feeds mainly at night by grazing in wet paddy-fields. Roosts in the daytime among mangrove trees or on rocks exposed during low tide. Affects fresh-water pools and marshes; tidal creeks and paddy-fields. Resident, from Andamans south to Greater Sunda islands, Moluccas, Australia, and New Zealand.

Sunda Teal (Oceanic Teal) *Anas gibberifrons* (Muller)
*JBNHS*, Vol. XII, No. 2, 1898. Painted by J.G. Keulemans.

Courtesy Vasant J. Sheth Memorial Foundation

Baer's Pochard *Aythya baeri* (Radde)
*JBNHS*, Vol. XII, No. 4, 1899. Painted by J.G. Keulemans.

Courtesy Tata Sons Limited

# GLEANINGS

Editors' Note: The pieces in this section have been selected from the Journal of the BNHS. The Journal has played an important role in the study of the natural history of the Indian subcontinent and a brief account of its history will not be amiss here. The following is extracted from an article in *JBNHS*, 1986, written by Dr. Sálim Ali for the Society's Centenary.

– A.S.K & B.F.C.

---

## The Journal

### Its Role in Indian Natural History

By 1886 – three years after the founding of the Bombay Natural History Society – the largely attended monthly meetings of its fast growing membership had become very popular but were tending to become more like social get-togethers than scientific seminars. To give the Society's activities meaningful significance it was considered desirable to publish a quarterly journal for maintaining a permanent record of the business transacted at the meetings – of the papers read and discussed and of the natural history specimens collected, exhibited and described by members, and hunting experiences of discerning sportsmen. Such a publication, it was felt, would also help to stimulate an intelligent and well-informed interest in Nature among the many who, though naturalists in the truest sense of the term, lacked a formal biological background. It would, moreover, keep the scattered outstation members in touch with the Society and with each other and encourage their participation in its activities. Up to that time there was no publication devoted to natural history in Bombay Presidency nor indeed in the Subcontinent as a whole. Little was known and recorded, and vast tracts of the country lay unexplored for their animals and plants. How richly the decision to publish a journal paid off is evident from the popularity and scientific prestige it has developed for itself and the Society over the years.

The first issue of the Journal, Vol. I (1), saw the light in January 1886 under the capable editorship of E.H. Aitken (EHA) who was the Honorary Secretary of the Society at the time. Another energetic naturalist-member, R.A. Sterndale, took over the editorship soon afterwards upon EHA leaving for England on home leave.

In the early days of the Journal, and until fairly recently – more or less all through the British period – the emphasis was largely on game animals and shikar. But the recorded experiences and field observations of well-informed and discerning sportsmen have helped to build up our knowledge of the life histories not only of quarry species but also others of less interest to the sportsman. A large proportion of the natural history of our game animals, both mammal and bird, has been acquired in this way, especially since the Journal made its appearance. Most of such knowledge is seminal and would have remained unavailable but for the published notes and articles of

< **Baer's Pochard**
*Aythya baeri*
Head and neck dark with greenish tinge contrasting with chestnut brown breast. A shy duck usually found in pairs or small parties, feeds mainly by diving, takes to air easily and is swift in flight. Regular winter visitor to large rivers and lakes of NE India and Bangladesh; also recorded from Bihar and Maharashtra (*JBNHS* XXVIII:1081). Globally threatened.

observant sportsmen. The latter consisted chiefly of British district officials, Army personnel and planters dispersed in remote backwoods lacking social amenities and congenial company, who had therefore taken to natural history and shikar by way of relaxation and recreation, a few of them developing into reputable authorities in their special subjects.

The second number of Vol. I, containing a heterogeneous variety of articles and notes on plants and animals – taxonomical, ecological and anecdotal – set the pattern which the rest of the volumes up to the present have more or less followed. The Miscellaneous Notes section which follows the main articles has always been the most popular feature with readers whose scientific interest is marginal; but many an anecdote casually recounted for its novelty for the writer has often proved to be meaningful to a scientist as corroborative or supplementary evidence for some pet theory of his own.

From chiefly shikar in the early days of the Society the accent in the Journal has steadily shifted to conservation on the growing realization that all was not well with our wildlife and that the once teeming game was vanishing fast throughout the country. Throughout its existence the Society has been deeply concerned about wildlife and environmental conservation, and the Journal has functioned as its main "mouthpiece" and an effective vehicle for its campaign against public and official apathy. All these destructive forces had to be countered by creating a healthy public opinion and pressurizing and persuading government to institute adequate legislative measures.

The special volume published in 1933 to commemorate the Jubilee year of the Society gives an excellent account of the Journal and its editors and functioning up to the 36th volume. These volumes represented the Society's contribution to the advance of our knowledge of the botany, zoology and nature conservation of the Subcontinent and adjoining countries. They point out how, apart from the results of scientific researches and field surveys, the Journal is unique in that it contains a vast amount of data – the notes and observations contributed by perceptive field naturalists which have helped significantly in promoting the refreshing trend of Indian biology from the museum to the field – from the study of the dead to the living: from taxonomy to ecology.

The need for protecting wildlife against unregulated hunting and large-scale commercial poaching by village shikaris was increasingly felt by forest officials and discerning sportsmen even since the early years of the 20th century. The wildlife protection movement visibly began in 1869 and culminated in The Wild Birds and Game Protection Act of 1887, shortly after the Journal was launched. But it was not until the Golden Jubilee of the Society in 1933 that wildlife preservation really came into sharp focus. It was the masterly address delivered by Mr. S.H. Prater, the Society's Curator on that occasion on "The Problems of Wildlife Protection in India" that seriously set the ball rolling and paved the way for the calling by the Viceroy (Lord Willingdon) – the Patron of the Society – of the all-India meeting at Delhi of prominent naturalists and sportsmen to review the deteriorating situation and suggest practical methods for effective conservation of wildlife. And it was the untiring and dedicated advocacy of the Society's stalwarts like Col. R.W. Burton, E.P. Gee and R.C. Morris, who through their authentic and well-researched articles in the Journal, kept the subject in sharp focus with government and the discerning public, leading to the establishment of most of the National Parks, Wildlife Sanctuaries and nature reserves that exist today, and to protective legislation culminating in the comprehensive Wildlife (Protection) Act of 1972. With proper implementation, this central legislation – itself based on the provincial Bombay Wild Birds and Wild Animals Preservation Act of 1951 (for which again BNHS was largely responsible) – should go a long way to saving what can still be saved of the splendid wealth and diversity of our once teeming wildlife and its natural habitats.

**"Rhino at Bay"**. Drawn by Thomas Williamson & Samuel Howitt. *Oriental Field Sports*, Vol. I, by Thomas Williamson, 1808.

It had become an unwritten convention for the Honorary Secretary of the time to be the editor of the Journal; though in later days after the appointment of the stipendiary Curator (N.B. Kinnear) in 1907, most of the actual editing and donkey work connected with the publication fell to the lot of the professional Curator. The Curator at the completion of Vol. 36 and for the next 15 years till he retired in 1948 to settle down in the U.K., was Mr. S.H. Prater. It is no exaggeration to say that the Journal reached the peak of its reputation and credibility during his editorship, especially while associated with Sir Reginald Spence as Honorary Secretary. Spence was an influential and dynamic personality and took a more active part in the affairs of the Society and in editing the Journal than most others. As executive editor Prater's name had become synonymous with the Bombay Natural History Society and he is largely responsible for the international recognition the Journal has acquired as the foremost natural history publication in Asia.

SÁLIM ALI

*JBNHS*, Vol. LXXXIII (Supplement).

**Red-headed Bunting** *Emberiza bruniceps* Brandt
*Birds of Asia*, Vol. IV, Parts XIX–XXIV, by John Gould, 1867–72. Painted by John Gould & Henry C. Richter.

In Memory of Jashbhai Javerbhai Patel of Village Valvod, District Kaira, Gujarat,
from Surendra Jashbhai Patel & Prabha Surendra Patel

# Red Ants' Nests

Some time ago I gave the members of the Society some account of the ways of the red ant (or yellow ant, as some prefer to call it,) known to formicologists as *Œcopylla smaragdina*. I did not then know how it constructs its curious leaf nests, so bitterly familiar to many of us. How I could live so long without knowing this, I cannot now explain, but in case there are others as stupid as myself I will describe the process. I first saw it going on in a tree with very large, leathery leaves, two of which were then being drawn together. Beginning at the point where the leaves were nearest each other, several ants laid hold of one with their jaws, and of the other with their hind feet, and began to pull as ants can. Further on, where the distance was greater, one ant seized one leaf with its jaws, then a second seized the first by the "small of the back," grasped the other leaf with its hind feet, and pulled. Further on still a chain of three, four, five, or even six, ants united the two leaves. As every member of the community which could find room for jaw or foot joined in, the space between the leaves was spanned by a web of ant fabric, in a state of the highest tension, very like the elastic in a "springside" boot. In the meantime a number of single ants were busy securing the labours of the rest with strong cords of silk, and tightening these as the leaves were drawn nearer and nearer. When a sufficient number of leaves have thus been bound together at their edges, the whole is made weatherproof with sheet silk, and divided into chambers and passages with the same material.

E.H. AITKEN

Karwar, 3rd November 1890.
*JBNHS*, Vol. V.

< **Red-headed Bunting**
*Emberiza bruniceps*
Male, a bright golden sparrow-like bird with rusty golden-brown head, throat, and breast. Tail longer than sparrow's and forked. Female ashy brown above, pale brownish below and without the red head. Seen in large flocks about cultivation which do considerable damage to jowar and bajra fields. Dr. Sálim Ali writes, "when settled on the green trees surrounding the fields they look like a profusion of bright yellow flowers in the distance." Winter visitor to Pakistan and greater part of continental and peninsular India. Breeds in Baluchistan, Afghanistan, etc.

EDWARD HAMILTON AITKEN or "EHA" (1851–1909), the son of a Scottish missionary, was born in Satara and educated in Bombay. He served first in the Education Department and later in the Customs. He was one of the eight original founders of the Society in 1883 and co-editor with R.A. Sterndale of the earliest issues of the *Journal*. He was interested in all branches of Zoology, but particularly in birds and insects, chiefly butterflies. For many years he was in charge of the Society's entomological section. He was a shrewd observer of humanity too, as his book *Behind the Bungalow* testifies. His first literary venture was entitled *The Tribes on My Frontier* describing the animals ordinarily met with in and around an Indian bungalow. Another popular book of his is *The Common Birds of Bombay*, and a less known book *The Naturalist on the Prowl* was written when he was in Kanara and "full of the scent of the jungles". All his books contain accurate knowledge pleasantly imparted with quaint humour and the joyousness of living which express so well the nature of the writer. Aitken was the first Honorary Secretary of the Society and continued as such till his departure from Bombay. He retired to Edinburgh in 1906 and died a few years later of Bright's disease.

The late Sir Norman Kinnear related to Dr. Sálim Ali an amusing story connected with EHA's death. The parish newspaper of the obscure little Scottish village to which EHA's father had belonged, thought it its duty to publish a fitting obituary of so distinguished a son of the village. The editor had probably never heard of the man or his interests and writings until his death got reported in outside newspapers. However, not to be outdone by his city confreres he ingeniously added to EHA's laudatory qualifications: "Mr. Aitken who had lived all his life in India was an expert on Frontier Tribes and Bungalow economy."

From "BOMBAY NATURAL HISTORY SOCIETY,
the founders, the builders and the guardians",
Part I by Dr. Sálim Ali which appeared in
*JBNHS*, Vol. LXXV, No. 3,
December 1978.

# Panthers Tree'd by Wild Dogs

On the morning of the 25th March, as my friend C– and I were shooting small game along the bank of the Gogra river in the Neelghal, Berar, a native shouted, "Bagh hai, Sahib; Bagh hai;" so we went up to him. In a bend of the river, in a tree on a very high bank on the opposite side, was something black, and there were animals moving below.

Binoculars at once cleared the vision. There were two panthers in a "Sallai" tree, one above the other, with a large pack of 10 or 12 couple of jungle dogs moving about below.

The upper panther was resting across a branch, and the lower one holding on perpendicularly. The difficulty was to approach. It was arranged that C– should go above and have the shot, while I went below. After a bit the lower panther made a jump, pursued by the pack in my direction on the bank, but he broke up a ravine. Just then C– shot the other panther dead, but he stuck in a lower fork when he fell. Some of the pack immediately came back and could be seen standing on their hind legs and licking the blood as it streamed from the beast out of reach.

My friend C– would have shot two dogs, but he had a miss-fire. I only got two or three long shots at the dogs. The panther shot was a fine male about 7' in length. The dogs made off, and we could not find the other panther.

Our informers said they saw the panthers tree'd at sunrise, and it was about 8 o'clock when we got there.

FRED WRIGHT

Chickalda, Berars, April 1890.
*JBNHS*, Vol. V.

**Silver-eared Mesia >**
*Leiothrix argentauris*

A brightly coloured arboreal babbler with black crown and moustachial stripes and silvery ear-patches. Forehead yellow, throat and breast bright orange-yellow. Wing with yellow edge and crimson patch, Tail coverts crimson. Resident, Himalayas from Garhwal eastwards to W. Bengal (Kurseong, Darjiling), Sikkim to Arunachal Pradesh, south to Assam, Nagaland, Meghalaya, Mizoram; Bangladesh (Chittagong hills). Dr. Sálim Ali in *Birds of Sikkim* writes that the species extends through Burma (Myanmar), Thailand, Malaysia, Yunnan, Indonesia, Indochina. Affects scrub jungle, from foothills up to c. 2,100 m. Seen in small parties of 6 or flocks up to 30 or more searching actively for insects among the foliage.

"**A Race for Life**", wild dogs chasing a nilgai. Drawn by Robert Armitage Sterndale. *Denizens of the Jungles*, 1886.

Silver-eared Mesia (Silver-eared Leiothrix) *Leiothrix argentauris* (Hodgson)
*Birds of Asia*, Vol. III, Parts XIII–XVIII, by John Gould, 1861–66. Painted by John Gould & Henry C. Richter.

Courtesy Bharat Floorings & Tiles (Mumbai) Pvt. Ltd.

# How the Monitor or Ghorapad (*Varanus bengalensis*) Defends Itself

To-day, whilst reading in the verandah I heard an unusual sort of noise, as of some creature careering over the gravel, and immediately got up to see what it was. A terrier, who had been asleep on the verandah steps, had also been disturbed by the noise, and when I looked up I found him standing face to face with a Ghorapad, or Monitor Lizard, about 3 feet long. They both appeared to be much astonished at the other's appearance. The Ghorapad evidently came to the conclusion that if there was to be a row in such an open space it ought to be fought out at once, and prepared himself accordingly, arching his back, swelling out the pouch under his throat, darting out his tongue in snake-like fashion, and hissing furiously. The dog for some time did not know what to make of such a strange creature, but eventually came to the conclusion that it ought to be worried and killed. He commenced the attack by rushing at his opponent's head, but the big lizard was equal to the occasion, and by suddenly turning round, presented his tail to the enemy, lashing out furiously with it and sending the gravel flying in all directions. Two or three times the dog returned to the attack, but always to find a tail where the head ought to be. Meanwhile a patiwala, hearing the noise, came on the scene, but quickly disappeared muttering something to himself of which only the words distinctly heard were "Karna ki waste". He shortly reappeared with a broad grin on his face and a thick blanket in his hand which he carefully threw over the Ghorapad, but the active creature slipped from under the *cumbli* and scuttled off for dear life towards the flower beds into which he escaped, thus saving his skin from adorning the family tom-tom, and depriving Gopal of a most tasty dish.

GEORGE K. WASEY

Marmagoa, 8th October 1891.
*JBNHS*, Vol. VI.


# Taming a Heron

One day during the recent monsoon a young Heron (*Butorides striatus*, the Little Green Heron) with a greenish-brown neck and body, white tipped wings, and green legs, flew into the verandah of my house, apparently in search of food. I caught it and for about ten days kept it under a large basket, feeding it with raw meat. I then gave it its liberty, but it refused to leave. It grew very tame and would feed out of my hand. Occasionally it would indulge in a bath in one of the dog's tins, and afterwards sit on a chair in the verandah. In the evening it flew away to roost in one of the large neem trees in the compound. It showed no fear of any of my dogs, and would give any of them who came too near a vigorous dig with its long bill. It remained with me for about six weeks, when as my Regiment was under orders to march, and I was afraid if left behind it would meet with an untimely end, I carried it down to the River Banas about two miles off and left it there.

W.S. HORE, LIEUT. COLONEL

Deesa, September 1891.
*JBNHS*, Vol. VI.

"**Monkeys at Cubbeer-Burr**". Drawn from nature by James Forbes, 1783. *Oriental Memoirs*, Vol. I, 1812.

## The Sagacity of the Langoor

The following story was related to me by Ballaji, Patel of Kusba Serazgaon, yesterday; when I was encamped there:–

"About a year ago," he said, pointing to a well in a garden by the roadside, surrounded with high tamarind trees, "a woman of the Mali caste left her baby 3 months old in a swing by the well, and went into the garden to work. After a while she returned and found the child gone. It had on a red garment. Search was made everywhere, and at last a monkey was noticed at the top of one of the high tamarind trees with something in its arms, and there was no doubt it was the child. They did not dare to do anything, but went off to a distance from the tree and watched. After some time the monkey came down and put the child back in the swing. It was found unhurt and the little boy is alive now."

The monkey was a female, and belonged to the common species known as "Langoor."

FRED WRIGHT

Elichpur, 6th November 1891.
*JBNHS*, Vol. VI.

TEXT AND ILLUSTRATION COURTESY VIDHI & SHAIVI KAPADIA (AGES 13 AND 9), MAAHIR & VEER SHAH (AGES 7 AND 2½), AND SHAAN MEHTA (AGE 2)

## An Aggressive Cobra

On the 28th June last a number of the convicts of the Kolhapore Jail were employed in cleaning up the compound of the State Hospital, and the sepoy in charge, Husain Bux, sat, watching the party, on the flight of nine broad stone steps which leads from the corner of the hospital compound up to the quarters of Mr. McGill, the Darbar

Veterinary Officer. It was about mid-day and the sepoy was sitting at the end of one of the steps, half way up the flight, with the entrance to Mr. McGill's little garden above and rather behind him to the left, when he felt a sharp smack, as from a flat object, on his back, just above his waist-belt, and, as he says, thought at the moment that some one had thrown something at him. Luckily, indeed, he did not put his hand behind to feel what the object was, for on looking round under his arm, without shifting his seat, he was horrified – and no wonder – to see a large cobra on the same steps just behind him, with hood expanded and ready to strike again.

Sidling off on to the ground, he shouted to his convicts, who with others ran up and pursued the snake, which now ascended the steps into the garden and took refuge behind some flower pots in Mr. McGill's compound.

Calming down from his fright, Husain Bux was for leaving it alone, saying that as Allah had spared him, so he would spare the snake, and no doubt, though a Musalman, he had something of the Hindu superstitious belief in the divinity of the cobra, and thought that such a peculiar visitation from the God would bring him luck. So, too, thought many natives, and when I talked over the occurrence afterwards with an old jail warder, he shook his head ominously and said, "He ought not to have killed that snake." However, killed the cobra was, and the men took it to Dr. G. Sinclair, who found it to measure 4 ft. 7 in. The snake, I fancy, must have been coming down from the garden above, when it saw the sepoy sitting on the steps, and that it should not have retreated or passed behind him, as there was plenty of room for it to do unnoticed, is curious, and such an instance of a cobra, when unalarmed, going out of its way to attack a man is perhaps worthy of record.

S.M. FRASER, I.C.S.

Kolhapore, 1893.
*JBNHS*, Vol. VIII.

## Wolf Cubs

I shall be glad to know whether the Indian wolf has ever been successfully trained to hunt with dogs, in a "bobbery" pack, as I hope to try the experiment before long.

Last month I came on the lair of a pair of wolves on the borders of the Runn of Cutch, containing 5 cubs (2 males and 3 females). I removed three of the cubs and left the others so as to tempt the parents back to their lair, hoping to be able to run them down with my greyhounds in the morning, but when I visited the place the next day before dawn, the old wolves had removed the two young ones during the night and had left their lair.

The three cubs, which I had taken, were successfully suckled by means of a village dog. Their foster-mother objected to them considerably at first, but soon became very fond of them, and will not leave them now. They are perfectly tame, come to call, and lick my hand, but they appear to be rather stupid, although exceedingly keen and plucky. All my dogs have taken a great fancy to the young wolves and play with them, but the wild parentage of the little cubs is especially noticeable at meal time when they bolt their food in the most ravenous manner imaginable. When they were only about three weeks old, they showed their hunting instinct by flying at the throat of a tame Chinkara fawn. Should they become unmanageable and dangerous, I shall send them to the Zoological Gardens in Bombay, but I still hope to train them to

**Red-rumped Swallow >**
*Hirundo daurica*
Upper plumage glossy deep blue; below fulvous white finely streaked with dark brown. The chestnut half colour on the hindneck, deeply forked tail, and chestnut rump help in identification. In flight the red rump sometimes appears very pale. Sexes alike. Seen in pairs or small parties in the neighbourhood of ruins, old temples, mosques, natural caves on hillsides, rock-overhangs, etc. their nest is a remarkable "retort-shaped" structure of fine mud pellets collected by the bird a mouthful at a time from the edges of puddles, and takes several weeks to build. It is always built under rocks, culverts, or bridges, or on house or verandah ceilings. The entrance is a narrow tubular passage which leads into a hemispherical chamber. Distribution: more or less over entire Indian plains.

Streak-throated Swallow (Indian Cliff Swallow) *Hirundo fluvicola* (Blyth)
*Birds of Asia*, Vol. IV, Parts XIX–XXIV, by John Gould, 1867–72. Painted by John Gould & Henry C. Richter.

Courtesy Jitu Bhansali, Bhansali & Co.

but whenever they lost sight of each other, while wandering from room to room, they became much disturbed – constantly making the shrill bird-like chirrips until they met. On bringing them into the Headquarter Station of the district, it was noticeable during the few days I spent there, that they had a strong antipathy towards all dogs, large or small; they would go for them at once, "spitting" and bouncing towards them; so much was this the case, that on several occasions I was afraid they would come to grief. I also remarked that they were hostile to small native children, but did not appear to dislike white children. On the other hand they were alarmed at the sight of a quarter-grown leopard cub, which was led about on a chain. Every one was much interested in their curious appearance and ways: long grey woolly hair, without spots, covered the head, back, tail, and came half-way down their bodies, giving them the appearance of wearing greatcoats, the remainder of the body and legs were covered with short grey hair sprinkled over with single black spots. On taking them out to play in the compound of the Travellers' bungalow at Damergaon, while waiting for a train, all the crows about the place made a very noisy demonstration on seeing them, collecting in large numbers on the trees overhead, cawing loudly and following them about – these crows appeared instinctively to know that the cubs, although so very small, were not ordinary cat kittens. Shortly after arriving at Dharwar the smaller cub became very ill and extremely weak, I separated it from its brother, who was inclined to pull it about, and for warmth used to keep it folded in my arms with a hotwater bottle; many a time I thought it was dying, but, no matter how bad it appeared – if it heard the chirrip call from its brother, it would at once struggle into vitality, tumble down, drag itself across the floor to where he was, and then lie down quiet and contented beside him; after a lot of careful nursing it gradually recovered. When about a month old they seemed not to care for their milk diet, so I tried beef tea in place, which they took most readily, lapping it up from the first. I possess two dogs, one an old spaniel, and the other a young half-bred dog; on first arrival at my bungalow the cubs did not at all like these dogs, but at once bounced at them and were most disagreeable and aggressive, the older dog would growl a good deal and avoid the charges by jumping on to a chair, while the younger one would bound round barking in play, but the cubs were very serious in their feelings, and I had great difficulties; however, after a time I got them to make friends with the younger dog, and finally with the old spaniel. Since then they have remained on the very closest and best terms, and the younger dog and cubs would have great play together round the compound. Often after lapping up their soup they would walk to the dogs and get their faces licked over by them, using the dogs as a napkin, for, until nearly seven months old, they never cleaned their faces by means of their paws, as cat kittens do, but, when the dogs were not present, would lick each other's face. Though the cubs were so extremely friendly with the two dogs – constantly lying down and going to sleep beside them – they would go for any other dog which came into the compound, and would also join the dogs in a race after one. While at dinner one evening a neighbour called to see me; it was dark and he was attended by a servant carrying a lantern, and on hearing the sound of footsteps approaching the bungalow my dogs rushed out of the dining-room barking; both the cubs were in the room at the time, and one of them bounded out with the dogs, evidently joining them in their hostile exhibition. I much fear this incident interested me more than it did the late caller, who was naturally rather taken back by the sudden rush of two dogs accompanied by leopards, although like the dogs the cubs were perfectly friendly when he came inside the house.

On the first occasion the cubs saw goats, when little more than a month old, they showed much excitement and hostility towards them, and one day when a goat happened to be tied in the compound, they got out of their enclosure, rushed at once at it, one catching hold at the shoulder and the other behind, but their teeth were so short they were easily shaken off. After the cubs commenced a purely meat diet – of raw and cooked meat when some three months old – a strong instinctive activity was noticed in their licking and eating moist earth – especially that of a red colour, and this they would do several times a day in a most persistent manner. It would be interesting to know whether this has been observed by those who may have brought up young tigers and leopards – if so, it would seem to point to the necessity of providing some fresh earth for these animals when confined in cages. The cubs are bad at tree-climbing, and very seldom in play go up on even a low tree, and when on one are almost as clumsy as a dog. I have never noticed them sharpen their claws against any tree; their power of sight is splendid, but their sense of smell is extremely deficient. When chasing each other, or one of the dogs in play, they always attempt, when going at full speed, to upset the play-fellow by striking his hind legs from under him, and when he falls, which he usually does, a spring is at once made at the neck, and he is then worried to an imaginary death.

G.S. RODON, MAJOR

Dharwar, 17th August 1897.
*JBNHS*, Vol. XI.

EDITORS' NOTE: The last three cheetahs in India were shot by a "Sportsman" in the winter of 1947. – A.S.K. & B.F.C.

IN MEMORY OF SHRI PRASANJIT S. KOTHARI OF PALANPUR.
ARTICLE AND PHOTOGRAPHS COURTESY DR AJAY P. KOTHARI,
PRESIDENT, ASTROX CORPORATION, COLLEGE PARK, MARYLAND, USA

# A Centipede Eating a Snake

I believe it to be a fact that Centipedes in general are eminently raptorial in their habits, attacking anything that they can overpower. It never occurred to me, however, that a Centipede would be bold enough to attack a snake as appears to be the case in this instance. Considering the subject worthy of record, and possibly of interest to members, I had a photograph taken about half the real size, depicting the Centipede and snake in the position I saw them and in which I understand they were found.

The specimens were received by Dr. Pedley, who sent them home (unfortunately

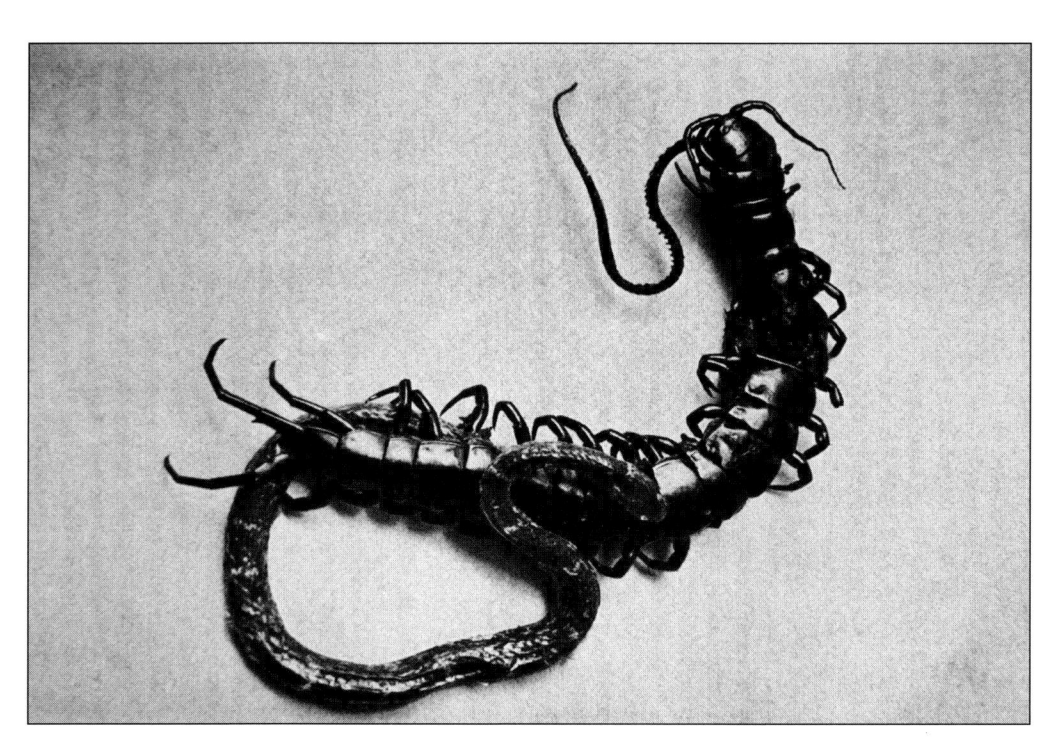

### Yellow-billed Blue Magpie >
*Urocissa flavirostris*

An attractive long-tailed purple-blue bird with black head, neck, and breast and pale primrose-yellow under-parts. The paler race in Western Himalayas, *Urocissa f. cucullata* (Gould) has white under-parts. Tail graduated, tipped with black and white, ending in long arching streamers that trail behind in flight. A small white patch on nape. Bill yellow, legs bright orange-yellow. Sexes alike. The Blue Magpies are arboreal birds usually met with in heavy jungle areas, but they also venture out into the trees among cultivation, and at times onto bare mountainsides at high elevations. They live in noisy parties of seven or eight birds. Parties fly in follow-my-leader style, tail spread and streamers waving behind. Call: Harsh cracking notes and sharp squealing whistles and at times mimic the calls of other birds. Resident with summer–winter altitudinal movements, normally between 2,000 and 3,300 m in summer; down to 1,000 m in winter. The Yellow-billed Magpies are found throughout the Himalayas from Hazara to Brahmaputra. They are divided into two races, *U. f. cucullata* (Western Himalayas) and *U. f. flavirostris* (Eastern Himalayas).

**Yellow-billed Blue Magpie** *Urocissa flavirostris* (Blyth)
*Birds of Asia*, Vol. III, Parts XIII–XVIII, by John Gould, 1861–66. Painted by John Gould & Henry C. Richter.

Courtesy Ashok Kumar Mehra & Co., Chartered Accountants

before they were identified) to his son for the Malborough College Museum, and it is to him I am indebted for the following information.

They were found on the floor of a house at Kokine, a suburb of Rangoon, the snake alive and writhing in the clutches of the Centipede. They were killed and at once transferred to a jar of spirits, and the owner thinking the incident an unusual one sent the specimens to Dr. Pedley. An inspection of the photo will show that the skin and flesh for about two inches has been completely removed from the tail of the snake and presumably eaten by the Centipede, which was one of the common large brown unstriped variety often met with in this part of the province.

It would have been very interesting to know if the snake was uninjured before the Centipede attacked it, but in any case as the snake was alive when found, it might still have held its own against the Centipede.

W.P. OKEDEN

Rangoon, January 1903.
*JBNHS*, Vol. XV.

# A Porcupine-Tiger Tragedy

I was in camp in the Nizam's Dominions two hot weathers ago with Mr. Mackenzie, Chief Engineer, when one morning a "Gond" trotted in to say there was a dead tiger lying in his field, done to death by a porcupine! I galloped out at once to the village about 8 miles, and was conducted by the villagers to the site of the tragedy. In a field as bare as one's hand lay peacefully dead on his side an almost full-grown well-conditioned tiger, with five large porcupine-quills stuck in his chest, like hatpins in a pin cushion. One hundred and twenty-eight paces distant were the remains of a large porcupine: only quills unfortunately, the body had been roasted and eaten before I could get there! But I was convinced after careful enquiry that practically the whole carcase had been found intact, minus one mouthful bitten out by the tiger. It was late evening before I got the dead tiger into camp, where we made a careful *post-*

**"Porcupine"** by Coleman. *Routledge's Picture Natural History* by the Rev. J.G. Wood, engraved by the Dalziel brothers, 1885.

**Common Green Magpie >**
*Cissa chinensis*

This bright leaf-green, long-tailed bird with cinnamon-red wings has a black band running backward through the eyes to meet on the nape. The green colour has a tendency to bleach to pale blue and the chestnut wings can fade to olive-brown. Red bill and legs. Sexes alike. Call: a long discordant quick repeated *peep-peep* or *ki-wee*; a raucous mewing note; rich melodious squealing whistles, and mimicry of other birds' calls (Sálim Ali, *Birds of the Eastern Himalayas*). This lovely bird is found below 1,500 m in the forests along the lower Himalayas from Uttaranchal east to Arunachal Pradesh, NE India, and Bangladesh.

**Common Green Magpie** *Cissa chinensis* (Boddaert)
*Birds of Asia*, Vol. II, Parts VII–XII, by John Gould, 1855–60. Painted by John Gould & Henry C. Richter.

In Memory of Bhanumati Rasiklal Kapadia & Chandrika Rajnikant Mody,
from Kapadia & Mody Families

*mortem* examination, regretting the absence of a doctor, capable of making a correct diagnosis of the cause of death. We found the pericardium bruised and discoloured, but as far as we could see the heart had not been penetrated by the quills, and was in a normal condition. On the other hand the liver and lungs were in shreds and looked like a black sponge in fragments. Outwardly the body was in excellent condition. It is not uncommon to find bits of quills in tiger's forepaws, pointing to the probability of their usually rendering porcupines harmless first with their paws, before going into them. We concluded that in this case, the tiger, being young and inexperienced, had jumped at the porcupine, and the large dorsal quills had driven with the impact into his chest. We found no bits of quills inside him, although the discoloured and disintegrated condition of his internal economy appeared to point to the probability of his having swallowed something fatal.

What was the cause of death? Perhaps some doctor-shikari will enlighten us. I have heard of tigers having been shot in an emaciated condition, and bits of quills having been found inside them; but this tiger was in splendid condition and certainly died very suddenly, because, firstly, he had left the kill practically uneaten; secondly, he had been able to run only one hundred and twenty-eight paces before death overtook him; and, thirdly, the fact of his having died in a bare field proved a sudden end, because, on the approach of death, if there is time, such animals invariably reach cover to die in.

G.E.C. Wakefield

Hyderabad, Deccan, 31st August 1913.
*JBNHS*, Vol. XXII.

# Habits of the Small Indian Mungoose

I am sending you a few notes on the breeding of the small Indian mungoose (*Herpestes auropunctatus*). I have a tame female of this species, which was given to me in August 1911, when about three weeks to a month old. On the evening of the 4th July 1912, she was observed pairing with a wild male, and gave birth to three young ones, 1 male and 2 females, on the night of the 22nd–23rd August – a period of 7 weeks.

On the 14th April of the current year, she again gave birth to two young ones, both females, and again on the 9th July she has given birth to two more, both females.

This mungoose makes a charming pet, and a most affectionate one, except when about to pair, when she becomes quite savage, and her temper is also most uncertain for a week before the birth of her young, and a month after. The extraordinary thing is, though she has absolute trust in my wife and self, yet she occasionally turns on us without rhyme or reason during these times. As for the servants and strangers, she will not allow them anywhere near her young, but allows us to handle them. She is a most restless mother for the first two or three weeks, constantly moving her offspring from place to place, carrying them as a cat carries her kittens.

The young when born are remarkably ugly, being practically hairless, and of a dark mouse colour. The eyes open on the 16th to 17th day. While sucking they purr like a cat and to this day when the mother drinks milk she purrs. We have never heard her do this at any other time, or over any other food. When angry they growl and spit. The mungoose, when attacked, only thinks of itself, never combining for mutual protection. But while the young are helpless, the mother is savage in their defence. There is no doubt that the sexes separate after pairing. I have an idea that the male would make a meal of the young, if he got a chance to do so. Two or three years ago I witnessed a strange thing. A cat of mine had just had her kittens, when a mungoose, owned by a neighbour, came into the room, and in a moment he had one of these kittens and made off with it. He was so quick that I had no time to rescue the poor little mite.

This mungoose is certainly a most useful pet, as all creeping things, such as snakes, centipedes, etc., are eaten by her, as well as scorpions, beetles, wasps, hornets, lizards, rats and mice, and insects of all descriptions. The only thing she does not eat are ants, but white ants are eaten with relish.

Our present house had a bad reputation for harbouring snakes and scorpions, and now there is not one to be seen. There is no doubt that the mungoose is immune to scorpion sting.

J.E. POWELL

Ghazipur, U.P., 31st July 1913.
*JBNHS*, Vol. XXII.


# Mongoose v. Cobra

In the last Journal, mention is made of a fight between the above. This reminds me of an incident which I witnessed while riding from Sirohi to Anadra one day. My attention was drawn by some brown thing moving in a small *Cassia auriculata* bush. It proved to be a mongoose, *Herpestes edwardsi*, attacking a cobra, *Naja naja*. The cobra (fairly young, some 4 feet I should say) was lying in waves over the twigs, the mongoose was leaping up at it from below, the cobra making such plunges at him as his unstrategical position allowed. A forlorn babbler was hopping dismally about on the twigs all the time; so probably the cobra had been paying attentions to the babbler when the mongoose's arrival drove him up to higher ground. I brought my horse's head close to the bush, but neither combatant seemed to notice us. After perhaps half a minute of upward jumps of the mongoose at the snake's body and as many counters by the snake, the mongoose ran off into the long grass 10 yards off. The snake lowered his hood and slid downwards off the bush into the low grass, moving off in a line at right angles to that taken by the mongoose. He had not gone far when the long grass stirred, and the mongoose peeped out. The snake stopped and raised its head, suspecting danger, but did not expand its hood. For a second or two they

**"Mungoose and Cobra"**. *Records of Sport in Southern India*, by General Douglas Hamilton, 1892.

remained thus, when in a second the mongoose had sprung forwards, nipped the snake's head and dragged him off, back downwards, into the long grass.

J.H. SMITH

Bhuj, Cutch, 10th January 1914.
*JBNHS*, Vol. XXII.

## Catching a Cobra with Bare Hands

On the 24th September 1915, a cobra about two feet long was observed in a yard in one of the enclosures of the Ghazipur Opium Factory. As snakes of this species are objects of special veneration to Hindus, the labourers employed in the vicinity did not attempt to kill it. The news of the discovery of a cobra in this somewhat unusual place filtered to another department of the Factory in which a man named Ghisan Komhar works. This man is an adept at catching snakes with his naked hands. He was soon on the spot and effected the capture in a few minutes without any apparatus of any kind. He approached the snake and seizing it by the end of the tail gave it a sudden jerk, and then lifted it up by the tail from the ground. The cobra was unable to twist round and bite him. He carried it away to his department and borrowed a penknife from the Assistant in charge and slipping his hand up the body of the snake forced open its jaws and broke its fangs. He then wrapped up the now harmless cobra in a piece of cloth and handed it over to his father who is a watchman, to keep until he could take it away. I saw the cobra a few minutes after its capture. It was alive and vigorous and full of fury. The moment the cloth was removed it erected itself and expanded its hood on which was very distinctly visible the characteristic pair of spectacles. It swayed about and struck viciously at its captor hitting him on his bare chest and on his fingers but never inflicting a bite.

Some years ago Ghisan Komhar captured a cobra about four feet in length with his naked hands to my knowledge. This cobra had wounded a boy who inadvertently came near its hiding place but did not succeed in inflicting a proper bite. The boy recovered. The snake was in such a position that it could not raise its head to strike properly. Hence the boy's lucky escape.

Ghisan Komhar who lives in Mianpura, a ward of Ghazipur City, is an adept in the perilous art he practises, and has caught many cobras in the manner described above. He told me in reply to questions I put him that he would feed the little snake he had caught about once in eight days and that its food would be one or two small frogs or toads (*mendki*) on each occasion.

Ghisan's object in catching snakes is to supply snake-charmers who buy them off him.

G.A. LEVETT-YEATS

Ghazipur, U.P., 25th September 1915.
*JBNHS*, Vol. XXIV.

**Himalayan Flameback >**
*Dinopium shorii*

A large golden-backed woodpecker with crimson crown and crest, crimson on lower back and black tail. Hind-neck black, the black continued forward as a black stripe to behind eye; a broad white band down either side of neck. Black moustachial streaks continued as a double line down centre of throat, with the intervening space pale brown. Breast and underparts buffy-white streaked and scalloped with black. Female similar to male but has white-streaked black crest. Resident, Himalayas from Himachal Pradesh and N. Haryana east to Arunachal Pradesh, NE India, and Bangladesh, also locally in the hills of peninsular India.

## Capturing Tigers with Bird-lime

At page 493, Vol. XXV of our Journal, there is a note by Colonel Burton on the method of capturing tigers with hay smeared with "glue," in the days of the Emperor Akbar. He enquires whether the plan is practical.

Himalayan Flameback (Himalayan Golden-backed Three-toed Woodpecker)
*Dinopium shorii* (Vigors)
*A Century of Birds from the Himalaya Mountains*, by John Gould, 1832. Painted by Elizabeth Gould.

Courtesy Diwaliben Mohanlal Mehta Charitable Trust

Rufous-bellied Niltava (Beautiful Niltava) *Niltava sundara* (Hodgson)
*Birds of Asia*, Vol. I, Parts I–VI, by John Gould, 1850–54. Painted by John Gould & Henry C. Richter.

Courtesy Maize Products, Ahmedabad

Whether practical or not, the practice seems to have survived in the Central Provinces. In 1890–91, I was stationed at Sambalpur in the Chattisghar division of that province, and was told that the jungle people there were in the habit of getting at tigers by laying down leaves smeared with bird-lime on paths frequented by the tiger they were after. It was mostly in the hot weather they did this in the neighbourhood of water pools, but probably they did it round a kill as well.

They told me that the tiger was annoyed by the leaves sticking to his paws, and tried to rub the leaves off on his head. The leaves then stuck to the face in such numbers as to blind the tiger, which could with safety be approached and speared, while in this helpless state.

W.B. BANNERMAN, SURGEON-GENERAL, I.M.S.

Madras, 13th March 1918.
*JBNHS*, Vol. XXV.

< **Rufous-bellied Niltava**
*Niltava sundara*

Male, forehead black; crown, rump, shoulders, and a patch on each side of neck shining blue; sides of head and back purplish blue-black. Below, throat black; rest orange-rufous. Female olive-brown with an ochre tint; tail rufous; a pale eye-ring; foreneck white with small patch of blue on either side. Difficult to find as it keeps to a great extent to thick evergreen undergrowth and prefers damp spots. It frequents pine forests, where there are damp nullahs with plenty of undergrowth on the banks of the streams running through them. Song described as squeaky and grating rendered as *s-i—chuck*. Distribution: from Murree Hills (Pakistan) in north-west Himalayas to Arunachal Pradesh up to 2,300 m and south in the hills of NE states in summer; foothills and adjacent plains in winter.

## The Tiger and the Train

About a month ago a curious incident occurred on the G.I.P. [Central Railway] main line where it runs through the Satpuras near Asirgarh.

Some surfacemen walking along the permanent-way came on the end of a tiger's tail lying beside the rail. It had obviously been quite recently cut off by a passing train. An inspection of the grass on the adjacent bank showed that some animal had made off with difficulty and a few steps were sufficient to bring part of the tiger into view. The surfacemen considered a closer inspection would be imprudent, although on their way to the nearest village they stoutly maintained to themselves that the tiger was dead, and they were thus able to impress on the local Shikari the simplicity of firing a bullet into its carcase and claiming the reward which Government pays for the destruction of tigers. The Shikari and the surfacemen promptly returned to the spot and the former (no doubt with a reduced charge for economy's sake) fired at the tiger, which at once got up and mauled him. While this was taking place a keyman on the G.I.P., who was also present, ran in and split the tiger's skull with an axe. The unfortunate Shikari died subsequently of his wounds. The tiger was found to have been struck by a passing train in the hind quarters and badly damaged. As such an occurrence must be very unusual it may be of some interest to the readers of the Journal. It is difficult to conceive how such a cautious and active animal as a tiger could get caught by a train: it might be accounted for by the passing of two trains simultaneously, or again the tiger at the last moment may have thought the other side of the track afforded more cover and security. I have heard of several instances of leopards being killed by trains, but these animals are much less timid of man and all his works than tigers are.

A.A. DUNBAR BRANDER, I.F.S.

Khandwa, C.P., November 1918.
*JBNHS*, Vol. XXVI.

## A Bird Passenger on a P.&O. Liner

I sailed from Bombay (Ballard Pier) on the 28th February last on my way to England. At the time of embarking my attention was drawn to a Hoopoe sitting placidly on the stern deck, but little notice was taken of the incident as it appeared nothing very much out of the ordinary, but after a couple of days on the high seas the passengers were very surprised to find that the bird was still on board. This was more or less passed over by the belief that it probably was a very tame bird owned by one of the sailors. I however, was interested in the case and went about enquiring from the ship's officers, sailors and laskars concerning this bird and finally was led to believe that it was the first time such a bird had "embarked on their boat" and nobody could account for the incident.

When we arrived at Port Said and anchored for about 6 hours, I made it my special business to see if the bird was still on deck, but it was nowhere to be found, and I believed that it had gone away having seen land and being tired of the voyage; but this evidently was not the case, as three days later I again saw the Hoopoe occasionally flying far out to sea in a very strong wind and rain, following the boat all the time and settling on the main deck railings. Most, if not all the passengers were interested in its performance, as the bird appeared strong and quite fit – it was wonderful to see it skim over the waves sometimes about six inches above the water.

After we passed Gibralter I was disappointed in not finding the brave little Hoopoe again, however much I searched for it and I am of opinion that it had found a suitable landing ground somewhere in Spain, at any rate I hope it was nothing disastrous.

It makes me wonder as to how this bird obtained food and water while on board for over a fortnight, as being practically an insect eater there appeared to be no suitable food for it on board, moreover nobody seemed to have noticed it ever feeding even if it should have changed its diet. It all appears very clever, if not strange, for the bird to have adopted this mode of migrating and at a time of the year when the heat of India was many degrees higher than that of Spain.

We have heard of birds of passage but this may be called a bird of "free passage" as it came on board unbooked and uninvited, and seemed to have known that it was not catching the wrong boat.

The boat I had travelled by was the P.&O. S.S. "Assaye", hired by Government as a Military Transport.

**V.R. WRIGHT-NEVILLE**

London, 15th April, 1924.

[There are many instances of land birds taking refuge on ships at sea, in most instances such species have been blown out of their course by high winds and stormy weather. The present instance is possibly unique as the weather conditions obtaining in Bombay during the time of the vessel's departure could in no way have effected the Hoopoe's movements. The Common Hoopoe (*Upupa epops*) is found in summer throughout the Palearctic region including the Himalayas. It migrates in winter to Africa, Arabia and India. Four subspecies are found in the Indian subcontinent. – EDS.]

*JBNHS*, Vol. XXX.

# Python and Monitor

I was in the Mergui District of Lower Burma with a friend John D. and we were trying to reach the source of a tributary of the Little Tenasserim River in what is almost virgin jungle.

At the end of one day's march our porters had just deposited their loads and were cutting bamboos to run up the rough shelters we were using at nights, as we were travelling light without tents. Some of the men were across the small stream we were following, when there came a cry of "Mwe; Mwe; Mwe gyi", (snake, snake, a big snake!).

One naturally thought that they had put up some large snake which had gone off, but they insisted that it was still there, just on the bank of the stream, which here was a pool over waist deep. As it had not been frightened or disturbed apparently by all this noise I thought they had probably lighted upon a Hamadryad, as these are not uncommon in the south of Lower Burma, so put together a gun and went across.

From the junction of a shallow stream I saw on the bank, but almost hidden by undergrowth, the huge girth of what was apparently an immense snake, and, climbing the bank within a few feet of it, found it to be an ordinary python (*Python molurus* ?) lying gorged. From the size and shape of the "bulge" I took it to be a Gyi (Muntjac or Barking Deer), the four shoulder and hip angles being plainly visible and I called back to D. that it was a python with a Gyi in it, asking him to come and lend a hand in hauling it down into the shallow stream, as not one of the Siamese Shan coolies with us would touch it, even when it had been shot.

Having got it into shallow water I proceeded to cut open the belly down the ventral shields over the carcase within, with a Dah (the weapon between knife and sword used by all Burmans and Shans, etc., for every purpose). Expecting to come upon the reddish *hair* of a Gyi I was surprised to come upon a mottled scaly *skin*; and cutting further exposed a great clawed hand, whereupon I shouted back to D. who, like the porters "wasn't having any," that it wasn't a Gyi but a young Crocodile inside and this I took it to be until I came to the head when I found that it was a monster Monitor Lizard (*Varanus* sp.).

Measurements carefully taken later by both of us with a steel tape gave the length of the Monitor as 5-ft. 9-ins. with a girth of 27-ins., (this after deflation; it was a very great deal more before being punctured, as decomposition had set in and it was greatly bloated, hence the enormous size it first appeared). The python was in perfect lustrous condition, having apparently but very recently shed a skin, and measured 14-ft. 11-ins. in length.

"A Langur in the Coils of a Python". From a photograph by W.R. Woodrow. *JBNHS*, Vol. IX, No. 4, 1894.

Surgeon-Major Chinner from Belgaum wrote in the *Journal* that he and Woodrow were tiger shooting in the forests of Canara in April 1892 when they found the subject of the engraving lying among dead leaves, coiled around the langur. The python was killed by the writer, and measured 12'10".

Now two things strike me as curious here. Firstly, that a python should attack a monitor at all, heavily armed as it is with powerful, sharp claws and a comparatively formidable set of teeth, and secondly, that having attacked it, it should have been able to envelop its prey so rapidly that the brilliant new suit it was wearing was absolutely without a scratch or mark upon it.

One would have supposed that a fierce battle would have ensued on the first grasping of the lizard by the snake, and that the former would have torn and scratched the snake seriously especially considering the relative size of victor and vanquished, the latter seeming the more powerful by far of the two. But not a mark was visible upon the glossy iridescent new skin of the rock snake.

The body was swallowed head-first, the fore-arms being pressed close to the sides, the hind limbs being bent backwards along the tail. The head was partially digested, but this process had only just commenced, and all the parts of the lizard were intact and perfect, and *not a bone broken*! In all ordinary cases the crushing action of the python, when coiled about its prey and preparing it for swallowing, breaks bones freely, but this tough lizard was intact and after extraction the limbs soon returned to their normal positions, showing that even the articulation of the joints had not been affected.

Although all snakes are more or less ophiophagus under certain conditions, the choice of a tough-skinned, horny-backed, and powerfully armed victim like the monitor seems very strange in a locality where more normal foods abound; it was in dense evergreen jungle where fowl, pheasant, the smaller cats, and all kinds of birds are plentiful.

I have no records of size, but although in India I have seen these Varanidae up to what must have been possibly six feet in length, this one of 5-ft. 9-ins. is the largest I have ever seen or heard of in Burma.

W.R. COLERIDGE BEADON

Rangoon, 17th April 1924.

[From the environment in which it was taken it is possible that the Monitor referred to is the Water Monitor (*Varanus salvator*) which is found in Bengal, Ceylon, S. China, Burma, the Malay Peninsula and Archipelago. The reptile frequents marshy localities or is found on trees overhanging rivers and streams. It grows to 7 feet in length. Two other species of Monitor occur in Burma, *Varanus flavescens*, the Yellow Monitor and *Varanus nebulosus*, the Clouded Monitor. The Monitor lizards are commonly miscalled Iguanas by Europeans in India. The Iguanas are entirely American, with the exception of two genera found in Madagascar. The Monitors are old world lizards. The term monitor is of curious derivation and is the result of an etymological error. The Arabic term for this lizard is "Ouaran"; this has wrongly been interpreted as a warning lizard, hence the Latin name Monitor.

The Python referred to by Mr. Beadon from the lustrous condition of the skin must have recently sloughed. Under these conditions the reptile is usually very hungry. Pythons in captivity are always very active after the process and quite ready for a meal and there is no knowing what a hungry python will not account for. An individual in the Society's rooms swallowed a black partridge, a brother python that had already commenced swallowing the same black partridge and a piece of red blanket which was entangled amongst its coils. –EDS.]

*JBNHS*, Vol. XXX.

**EDITORS' NOTE** – From the evidence that the python had no injury marks, which would have resulted from a struggling monitor lizard, that the latter's ribs were not crushed, and that its body was bloated and decomposed, it is very likely that the python devoured a dead lizard. – A.S.K. & B.F.C.

**Long-tailed Broadbill >**
*Psarisomus dalhousiae*

An attractive grass-green, arboreal and social bird with blue in wings and tail, black head, and yellow throat, remarkable for the flat broad bill and the tail of narrow graduated feathers. Keeps in flocks of 15 to 30, moving about in the forest canopy in sprawling follow-my-leader style from tree to tree. Utters a distinct loud, sharp whistle *tseeay, tseeay* repeatedly 5 to 10 times usually while in flight. Distribution: Himalayas from Garhwal eastwards to Arunachal Pradesh, Manipur, Mizoram, and Bangladesh.

Long-tailed Broadbill *Psarisomus dalhousiae* (Jameson)
*Birds of Asia*, Vol. I, Parts I–VI, by John Gould, 1850–54. Painted by John Gould & Henry C. Richter.

Courtesy Raj Shekhar Parikh, Renaissance Diamond Corporation, New York, USA

## Python Attacking a Spaniel

Perhaps the following short account may interest you, in fact I think the occurrence is probably unique.

Whilst camped at Bomanballi I was out with my wife, Mr Wagle, foreman of Sambrani Sawmill, and my dog (a spaniel) on the 11th May forenoon, choosing a site for a tent. The dog was running ahead amongst some bushes when we heard terrible yells as though something was killing the dog. I rushed up with my walking stick and found the dog caught by a python. The python was coiled round the dog several times and all I could see of the dog was a small portion of its hind quarters. I beat the python as hard as I could with my stick and gradually it uncoiled itself until the dog was able to escape. The dog was streaming with blood from the mouth and chest but wanted to return and fight the python. The python measured 11¾ feet and had apparently been waiting patiently in the hope of catching a monkey as they were playing in a tree under which it was lying. The dog was bitten on the under lip and on the chest but no bones were broken and it was otherwise unhurt. The python was killed (or at least so we thought) and carried back to camp and left under a shady tree whilst we had breakfast prior to skinning it. Shortly afterwards a small boy ran into the tent and said the python was still alive. Sure enough it was not only alive but had very nearly succeeded in catching one of our chickens. It was the second python caught within a fortnight within a few hundred yards of our tent.

R.G. SMITH, CAPT., I.F.S.,
ASSTT. CONSERVATOR OF FORESTS

N.D. Kanara, Camp via Dharwar, 30th May 1924.

[In the "Snakes of Bombay Island and Salsette" published in Vol. XXX, No. 1, an incident is recorded of a python attacking a terrier. – EDS.]

*JBNHS*, Vol. XXX.

## Tiger Killed by a Cobra

The following is an account of a tigress which died as a result of its being bitten by a cobra in the Jamnagar Zoo, Kathiawar. A correspondent reporting the case in the *Times of India* writes as follows:–

"An incident which is probably unprecedented in the case of animals in captivity occurred recently in the Zoological Gardens at Jamnagar. Among the fine specimens, kept there by His Highness the Jam Saheb, was a tigress with three cubs. On going his rounds early one morning the keeper was not a little surprised to find the tigress stark and stiff. Immediate search for the cause of this sudden demise on the part of what had been overnight a thoroughly healthy animal, revealed a cobra sleeping peacefully coiled up in a corner of the cage. A gun speedily did the needful despatch. On examination the tigress was found to have been bitten in the cheek. The three cubs were all very much alive and well. How they failed to attract a bite is 'wrapt in mystery'. Or perhaps the cobra, having expended its venom on the tigress, may have bitten one and proved harmless. However, no mark of such a bite could be traced, and the three cubs are now as lively as crickets, the cynosure of many curious and admiring eyes. His Highness' guests are often at the Zoo handling these playful little beasts and watching them at feeding time. In this function a trio of sturdy goats plays an important and, it must be said, extraordinarily complaisant part. The foster mothers play the game in most sporting fashion and very seldom attempt to butt their sturdy 'offspring.' Doubtless the time is not far distant when sharper claws and teeth will

**Silver-breasted Broadbill >**
*Serilophus lunatus*

A sluggish arboreal ashy-grey bird with a short recumbent crest overhanging the nape. Lower back chestnut. Wings black with contrasting chestnut, blue, and white markings. Tail black, graduated, the outer feathers white-tipped. Prominent yellow patch around eye. Female similar, but with a demi-gorget of white-tipped feathers on each side of neck. Arboreal and crepuscular. Loose parties of 5 to 20 active during early morning and around dusk but also hunt lethargically during the day. Perches very erect, the tail kept well down and frequently twitched. Call: a soft musical whistle; a low *chir-r-r* uttered at rest and on the wing have been described; a low mouse-like squeaking when alarmed. Resident, Himalayas from Nepal east to Arunachal Pradesh, NE India, and Bangladesh. Duars, foothills, and up to c. 1,700 m in tropical semi-evergreen and evergreen biotope. Affects sal and mixed secondary tree and bamboo jungle.

Silver-breasted Broadbill (Hodgson's Broadbill, Nepal Collared Broadbill)
*Serilophus lunatus* (Hodgson)
*Birds of Asia*, Vol. I, Parts I–VI, by John Gould, 1850–54. Painted by John Gould & Henry C. Richter.

Courtesy Tolani Shipping Co. Ltd.

bring a change on the scene, but at present the trio of *budmashes* is thriving apace and looking very well upon its enforced diet."

The following extract from the report of the Veterinary Surgeon in charge of the Zoo was forwarded to us by the Private Secretary to H.H. the Maharaja of Jamnagar.

"The tigress *Sunder* was alright on the 9th December. Next morning when the keeper of the animals went to see them as usual, he found the tigress dead. He informed me, whereupon I went there. The tigress was taken out of the cage. On an examination it was found that she was bitten on the lower jaw just below the angle of the mouth on the left side. The part bitten was swollen and there was haemorrhage from the nostrils and from the rectum. On a post-mortem examination it was found that all the organs inside were little or more congested and blood was found coagulated on exposure to atmosphere, *i.e.*, after removal from the blood vessels."

"The tigress left three cubs – two females and one male about a month old. For their nourishment she-goats are kept and they are suckling them. The cubs are quite well and in good condition. The incident occurred during the night."

*JBNHS*, Vol. XXX.

# A Panther Shoot at Sea

The following rather curious case of a Panther being shot at sea may interest your readers.

In February last a small *Pattamar*, of about 60 tons, anchored off the Port of Mangalore and sent ashore two of the crew to report to the Port Officer, Mr. Sims, that they had that morning, been attacked by a "Tiger" on board. One man had been scragged down the face and chest and the other had half his scalp removed.

As Mr. Sims asked me to accompany him out to the ship I witnessed what subsequently occurred. On arrival at the *Pattamar* we found she was five days out from a Port south of Bombay (Chaul) where she had loaded salt. They had not touched at any other port. That morning when a member of the crew had gone to the forepart of the ship (which is covered in for about 18 feet) for firewood, he was attacked by the "Tiger"; a friend who went to his help also got scalped.

We found the *Pattamar*, with the balance of the crew of six men still aboard, was loaded from keel to gunwale with bags of salt. There was a space forward where they kept firewood and in which we were told was a tiger. They had been at sea with it for five days and never knew it was there till that morning.

After some time we managed to get the panther to move, it crawled into the light and suddenly raised its head between two pieces of timber. Mr. Sims then shot it in the head, between ear and eye, with a .320 revolver.

It proved to be a full grown male panther, measuring between pegs 5 feet 9 inches.

In Canarese these panthers are called "Ni Kurukers," living chiefly on dogs and small game.

At the port where the salt was loaded the boats go up a creek and tie up alongside. The panther presumably got on board just before they put off and crawled into

**Satyr Tragopan >**
*Tragopan satyra*

Male, head black, blue facial skin and throat, and bluish lappets and horns which are erected in display. Rest of upper-parts crimson and brown, sprinkled with black-bordered white spots. Under-parts crimson, sprinkled with white ocelli as above. Female varies from rufous brown to ochre-brown, and is vermiculated, mottled and spotted with black and buff. Keeps to oak, deodar, and rhododendron forest with dense undergrowth and bamboo clumps, shrubberies on steep hillsides, and narrow ravines. Pairs or small parties may be seen scratching and feeding like junglefowl. Food: mostly ferns and other leafy vegetable matter. Call: a loud *wak* repeated several times, also *kya…kya…kya…* like the bleating of a goatkid. Resident, Himalayas from N. Uttaranchal to W. Nepal between 2,400 and 4,300 m, descending to lower levels in severe winters.

Satyr Tragopan (Crimson Tragopan, Crimson-horned Pheasant)
*Tragopan satyra* (Linnaeus)
*A Century of Birds from the Himalaya Mountains*, by John Gould, 1832. Painted by Elizabeth Gould.

Courtesy Mahendra Brothers

the first available lair, which was on top of the stock of firewood. As the stock got reduced, the man who went for it daily at length came within striking distance of the panther and was caught.

The two wounded men were in hospital for over a fortnight, but eventually pulled round.

N. KIRWAN

Mangalore, 4th August 1926.

[It would be interesting to learn what this panther was after, and what induced it literally to embark upon such a novel undertaking! An instance is related by Sir Samuel Baker of a tiger, during an inundation of the Brahmaputra, having climbed up in the night on the high rudder of a vessel much to the discomfiture of the native steersman when he beheld the visitor in the morning. In the uproar and confusion that followed the startling discovery, the tiger leaped from the rudder on to the barge which was lashed to the steamer, and having knocked over two men in its panic-stricken onset, bounded off the flat and sought security upon the deck of the steamer alongside, where he was eventually shot in the paddle-box. – EDS.]

*JBNHS*, Vol. XXXI.

## An Example of an Assisted Passage

On April 26, 1926, I sailed from Bombay in the S.S. *City of Exeter* which left about midday. About 3.30 p.m. I discovered that we had on board a single House-Crow (*Corvus splendens*) which had apparently failed to notice our departure in time. It looked rather bewildered at finding itself at sea, and grew more and more uneasy as dusk approached, launching out over the sea and then regaining the ship with difficulty. It was a dark dusky looking bird obviously of the typical race from Bombay, and not the pale *zugmayeri* which is the form at Karachi whence the *City of Exeter* had come.

The crow stayed with the ship for the next few days roosting on the rigging and receiving food and water by the good offices of the crew. We did not put in at Aden. We passed Perim on May 2 about 7–8 a.m. and after that I watched the bird making several attempts to fly towards the mainland which was visible. Its courage always failed however and it returned to the ship. By evening however it had disappeared and I was told that it had apparently flown on to another ship. Unfortunately this ship was steaming in the same direction as ourselves, or I might have told a good story about corvine sagacity in taking a ship back to India! As an observed example of how stragglers leave their habitat, this seems worth recording.

HUGH WHISTLER, F.L.S., F.Z.S.

Battle, Sussex, 8th August 1927.
*JBNHS*, Vol. XXXII.

HUGH WHISTLER was born in Lincolnshire on 28th September 1889. In 1909 he came to India and joined the Indian Police. His first station was Phillour, and he afterwards served at Rawalpindi, Ferozepur, Jhelum, Gujranwala, Ambala, Jhang, Kangra, and Simla; while on a short leave he visited Dalhousie and Kulu, as well as Kashmir. By the time he retired, Whistler had a very wide knowledge of the whole of Punjab and its avifauna. Wherever he went he noted and collected birds and the great store of knowledge he accumulated was communicated to the BNHS Journal and *Ibis*.

After the First World War, three of the Society's oldest members, W.S. Millard, Sir George Lowndes,

**Hunting of Blackbuck with Cheetah**. Drawn by James Forbes in South Gujarat. *Oriental Memoirs*, Vol. I, 1812.

The Mughal emperors kept large cheetah-khanas for tame hunting leopards which were used to hunt deer or antelope. A cheetah would be taken on hunting expeditions, chained and hooded on carts, and released after removing the hood or eyefold when the quarry was sighted. Capable of high speeds it would soon catch up with the animal, strike it down, and seize it by the throat. A ladleful of blood from the animal's throat would be offered to the cheetah to persuade it to release its hold. During the early part of the twentieth century cheetahs were still kept by the Maharajas of Bhavnagar and Kolhapur and the Nizam of Hyderabad.

and F.J. Mitchell, voiced the necessity of publication of a popular illustrated work on Indian birds and Whistler was approached to undertake the authorship. A happier selection could not have been made; *A Popular Handbook of Indian Birds* passed through four editions.

Early in 1925, Whistler married Margaret Joan, second daughter of Lord Ashton of Hyde. In April 1926 he took leave pending retirement and left India. Afterwards he visited India with Dr. Ticehurst and sometimes with his wife, travelling to remote places. He helped the Society in the survey of the birds of the Eastern Ghats and also of other regions. He died in England in early July 1943.

*JBNHS*, Vol. XLIV, No. 2.

# The Hunting Leopard in the Central Provinces

I shot a Hunting Leopard on December 26, 1926, in the Harrai Jagir (Chhindwara district) very near the Narsinghpur border, at a place called Kodari, lying on the main Narsinghpur-Chhindwara Road. The skin and head were mounted for me by Messrs. Van Ingen & Van Ingen, Mysore, who pronounced it to be *Acinonyx jubatus*. It is a splendid specimen.

Mr. C.F. Turner, C.I.E., I.C.S., Commissioner of this Division, suggests that I should report this to the Natural History Society for record, as he believes no wild hunting leopards have been known in these parts for the last 50 years.

J.M. RICHARDSON, I.M.D.

Narsingpur, 12th July 1929.

[Writing of the Hunting Leopard, Dunbar Brander (*Wild Animals of Central India*, p. 273) states that the animal has now almost entirely disappeared from the Central Provinces without apparent reason. He only knew of three animals being procured in the last 20 years. Rumours of their existence in parts of Berar, the Seoni Plateau and Saugor still persist, and the writer believes that one or two may still be found. In Vol. XXVII, p. 397 of the Society's Journal, Brig.-Genl. R.G. Burton gives several records of Hunting Leopards, shot in the Berar District between the years 1890 and 1895. They appear to have been fairly common in that district about the period mentioned. On the same page, the late Col. L.L. Fenton records the distribution of the Hunting Leopard in Kathiawar. In Vol. XXVI, p. 1041, there is an account of its occurrence in the Mirzapur District, U.P. Little is known of the present distribution of this animal in India. It is said to be still fairly common in the Hyderabad State, but we have no authentic records. – EDS.].

*JBNHS*, Vol. XXXIV.

TEXT AND ILLUSTRATION COURTESY NAVDEEP CHEMICALS PVT. LTD.

## How the Monitor Lizard Sits in its Burrow

In the last week of July this year, the Zoology staff and Post-graduate students of our college, accompanied by three snake-charmers, went out into the jungle round about Midhaku, a small village near Agra. The purpose was to see the fauna of this place in its natural haunts. Besides catching several centipedes, insects, snakes and other animals, we secured eight monitor lizards (*Varanus bengalensis*), which are quite common here. The snake-charmers had a remarkable sense of recognition of their traces. My attention was especially drawn to the bold way, in which, after digging a burrow waist-deep with the *kudal*, they would half dive into it and fearlessly drag out a big, struggling specimen by the tail. The monitor has strong jaws and can give a bite not easy to forget. Last year a snake-charmer was bitten, and after many futile efforts to release his finger from inside the reptile's mouth, we had, as a last resource, to cut the jaws open with scissors. It was a terrible bite! In spite of such experiences however, the snake-charmers do not seem to think much of risking their heads and hands in a monitor's burrow. On enquiring, I learnt that the creature has the habit of always sitting in the burrow with its tail nearer, and its head away from, the opening. Once the tail is caught, it cannot turn its clumsy length round to bite at the offender, and the catcher is safe.

BENI CHARAN MAHENDRA,
LECTURER IN ZOOLOGY

St. John's College, Agra, 26th October 1929.
*JBNHS*, Vol. XXXIV.

**Golden-fronted Leafbird >**
*Chloropsis aurifrons*
An active leaf-green arboreal bird with orange forehead, blue shoulder patches, and slightly curved black bill; around eye, ear patches, and lower throat black; chin and cheeks dark blue. Female paler and duller. Keeps in pairs or small parties to thick foliage of trees where its plumage blends perfectly well with the green leaves making it difficult to spot. Dr. Sálim Ali writes, "All chloropses are important flower-birds and responsible for pollinating the blossoms of numerous species of trees and shrubs in their quest for nectar." It is an accomplished mimic; very convincing imitations of various birds' calls are given in quick succession. Affects well wooded areas, forests as well as neighbourhood of habitation. Found throughout the Indian Union, Bangladesh, Sri Lanka, Myanmar.

## "Feline Government Gazette"

I have been recently transferred, (though not to a better land) and I was informed of my impending transfer first by a telegram from the Private Secretary to H.E. the Governor, then by a D.O. letter from the same official, and then by my appointment

Golden-fronted Leafbird (Gold-fronted Chloropsis) *Chloropsis aurifrons* (Temminck)
*Birds of Asia*, Vol. III, Parts XIII–XVIII, by John Gould, 1861–66. Painted by John Gould & Henry C. Richter.

Courtesy Dimexon Diamonds Limited

being notified in the *Bombay Government Gazette*. Finally I arrived here. Everything was done in due order.

But is any explanation available of how the transfers of tigers are arranged? In Khandesh, and presumably in other parts of India, the country is divided into districts, each occupied by a resident tiger. On the transfer of an incumbent to a better land, the vacancy is almost at once filled by his successor taking over charge.

How is the vacancy notified and who communicates it to the transferee? If there is a supply of *umedwar*, or candidate, tigers always mouching round on the chance of stumbling into a vacant post, one would presumably meet more tigers about than one does. Or is there a Central Tiger Reserve Depot or Training Battalion in Central India, from which a competent acting tiger is appointed to an acting vacancy with the possibility of being made permanent if suitable, or of losing it if shootable? If so, how is Headquarters notified of an incumbent's decease? Are transfers notified in the *Feline Government Gazette*, or in the *"Tigers of India"*?

I have no adequate explanation, and seriously I should like to know.

H.F. KNIGHT

Collector's Bungalow, Sholapur, 2nd January 1930.
*JBNHS*, Vol. XXXIV.

# Tigers Swimming

During several years of shooting in the "Sunderbunds" forests (Bengal), I discovered that tigers there readily take to water and in some instances swim considerable distances (3 or 4 miles), and that, in tidal rivers, with a 4 to 5 knot tide running during Spring tides, but what struck me most, however, was the intelligence displayed by tigers in choosing their time for swimming, which was invariably at or about high water, when they were able to "take off" and land on hard ground, and anyone who has had experience in the "Sunderbunds" will appreciate what this means! – At all other states of the tide one has to flounder up several yards of bank through more than knee-deep mud, which would prove very embarrassing to a heavy animal like a tiger.

Another curious feature about the "Sunderbund" Forests is the entire absence of fresh water (except in cultivated areas); – there are no fresh water ponds or streams, and tiger, deer, etc. have nothing but salt water to drink during the dry season, i.e. November–May, which seems hardly credible, but which nevertheless I am reliably informed is a fact.

Fresh water is however obtainable in the islands in certain suitable localities, but one has to dig 4 or 5 feet deep for it, and this is the only method of replenishing stocks of drinking water if one is making a prolonged stay in or around the uninhabited islands on the sea front.

These fresh water "holes" are soon discovered by the game who go mad after the water, and on one occasion I sat up over a hole to watch the game that came to drink, and spent one of the most enjoyable nights in the forest. Numberless cheetal, pig, jackals, and whilst I was dozing, a tigress and two half-grown cubs visited the hole, and the next morning I found tracks of a large monitor lizard and a python who had also quenched their thirst.

Tiger across a River
*Oriental Field Sports*, Vol. I, by Thomas Williamson, 1808. Drawn by Thomas Williamson & Samuel Howitt.

Courtesy Dilipkumar V. Lakhi

Orange-bellied Leafbird (Orange-bellied Chloropsis)
*Chloropsis hardwickii* Jardine & Selby
*Birds of Asia*, Vol. III, Parts XIII–XVIII, by John Gould, 1861–66. Painted by John Gould & Henry C. Richter.

Courtesy D. Navinchandra & Company

These holes have to be dug at a wide angle, i.e. about 45° (saucer shaped), as otherwise the walls collapse owing to the large amount of sand in the earth, and it is for this reason also that the holes soon "dry up" through the game pushing back the earth and sand in trekking up and down the sides or walls of the hole.

W.A. HICKIE

Singapore, 12th November 1929.
JBNHS, Vol. XXXIV.

### < Orange-bellied Leafbird
*Chloropsis hardwickii*

Male bright leaf-green above with a pale greenish blue shoulder-patch and dark purplish blue wings and tail. Bright brownish orange below, with breast, throat, chin and sides of head deep bluish black. A bright cobalt-blue moustachial streak. Female: almost entirely green, a pale blue shoulder patch; orange of under-parts paler and less extensive, paler blue moustachial streak. Tail green; no black on throat. Arboreal. Keeps in pairs or small parties in foliage canopy. Dr. Sálim Ali mentions in *Birds of Sikkim* that on one occasion he observed a gathering of over 50 feeding on a flush of Mahua (*Madhuca indica*) flowers in the company of sibias, spider-hunters, yellow-backed sunbirds, and white-eyes. It is a remarkably versatile songster and an excellent mimic. Common resident, found from c. 600 m to 2,500 m in Himalayas from W. Himachal Pradesh east to Arunachal Pradesh, NE India, and Bangladesh.

## Eleven Koel Eggs in a Crow's Nest

On 9th June I came upon a Common Crow's nest (*Corvus splendens*) at Bhandup (Salsette) containing eleven Koel (*Eudynamys scolopacea*) eggs belonging to two distinct types and apparently the product of two females. There were four eggs of one type and seven of the other. I removed one egg of each type which were kindly confirmed by the Society as having been laid probably by different females. The nest contained no crow eggs but bore obvious signs of having recently had an egg broken in it. As far as could be ascertained the locality certainly appeared to hold more koels than crows, a circumstance which may account for the concentration on this nest.

On the 17th I visited the place again to find the nest empty! There was another nest within 50 yards — not previously noticed — in which a crow was sitting.

HUMAYUN ABDULALI

Andheri, Salsette, 20th June 1931.
JBNHS, Vol. XXXV.

HUMAYUN ABDULALI (1914–2001) along with Sálim Ali was responsible for the rejuvenation and relaunching of the Society when India attained independence and the majority of the Society's supporters left India. They nursed the Society during this difficult period, and during the ten years that he was the Honorary Secretary, Humayun saw to it that the Society regained its solid foundation in the study of Natural History. Working with Sálim Ali, he ensured that the Society had the crucial support of the State and Central governments in that land and funds were made available for housing the Society and its valuable collections, and continued publication of the Journal, which was at that point of time the main activity of the Society.

Humayun was born in Kobe, Japan. The family returned to India and settled in Bombay in 1925. His association with Br. Navarro at St. Xavier's High School started his lifelong interest in Natural History.

Humayun's significant scientific contributions include cataloguing the collection of birds at the BNHS and the study of the bird fauna of the Andaman and Nicobar Islands. The nearly 300 papers he published reflect the wide scope of his interest in Indian Natural History. He made significant contributions to the conservation of India's wildlife heritage. The drafting of the Bombay Wild Bird and Wild Animals Protection Act in 1951, the basic source for the Wildlife Protection Act passed in 1972, the filing of the first Public Interest Litigation which saved the Borivli National Park in Bombay from being destroyed by a highway, and ban on the export of frogs' legs and junglefowl hackles are peaks in the conservation movement in the country, which owe their success to Humayun's efforts.

Extracted from a tribute by J.C. Daniel
published in *JBNHS*, Vol. C, Nos. 2 & 3,
Aug.–Dec. 2003.

# Remarkable Behaviour of a Tigress

While on the survey of the Eastern Ghats I was told a very strange tale of the behaviour of a tigress in the Nallamalli Hills. As a matter of fact the story has gone its "umteenth" round among the Officers and Rangers of the Forest Department in the Cuddapah and Kurnool districts.

The scene was in the Nallamalli Hills, at Iskakundam Bungalow, which is about 30 miles from Diguvametta. The Conservator and his Deputy had occupied the two rooms in the forest bungalow, the rest of the staff were in the out-houses.

The Deputy was writing his report under a petrol lamp in his room when he felt something brush against his chair. He slightly turned his head and saw a tiger rubbing itself along the back of the chair!

I wonder what a good many of us would have done? Shouted? Screamed? Jumped on the table? or fallen down in a dead faint?

The Deputy did none of these things. He was not a big game hunter. I doubt if he had ever killed a thing in his life. He calmly got up from his chair and as calmly walked out of the room – closing the door behind him.

As calmly he announced to the Conservator in the next room, that he had securely locked a tiger in his room.

None of us can blame the Conservator for jumping to a quite natural, though in this instance, unwarranted conclusion. The Deputy never drank a drop of spirits in his life.

They both went outside and peeping through the barred window saw the tigress placidly rubbing herself against the table. The first shot hit the animal but in leaping up she upset the lamp which fortunately was extinguished by the fall. The second shot was fired from the roof, through a hole in the thatch.

On examining the animal it was found that she suffered from a wound on the thigh, which was alive with maggots.

Now here is something for our big-game hunters to solve. Though we cannot possibly afford a prize for the best explanation put forth, it will give us an insight into the philosophy with which each big-game hunter and others approach the fascinating study of animal behaviour. What brought the tigress into the bungalow?

V.S. La Personne, m.b.o.u., Asst. Curator

Bombay, 31st March 1932.
*JBNHS*, Vol. XXXVI.

**Jerdon's Chloropsis >**
*Chloropsis cochinchinensis jerdoni*
An attractive arboreal bird, particularly fond of feeding on parasitic *Loranthus* flowers and also those of Silk Cotton, Coral, Palas, and other flowering trees. Bright green, a black throat-patch broken by a purplish blue moustachial streak in the male, a bluish green throat-patch in the female. In both sexes the throat-patch is bordered with yellow. Generally found in open country, the bird is partial to groves of trees around villages and cultivation. The food consists of fruit, seeds, insects, and nectar of various flowers. It lives in pairs which often join mixed hunting parties, and is a very active and restless bird. A particular characteristic of the chloropsis group is a remarkable proficiency in mimicry of other bird calls. Distribution: the Gangetic plain, peninsular India, Sri Lanka; the nominate race *cochinchinensis* in NE hill states, Bangladesh, Myanmar.

# A Newly Born Bison Calf

An interesting event took place while a party of us (Col. and Mrs. Newcomb, Mr. and Mrs. Sálim A. Ali, Mrs. Morris and myself) were watching a herd of bison.

One of the cows calved. Ten minutes after birth the calf was walking about and

Jerdon's Chloropsis (Gold-mantled Chloropsis, Blue-winged Leafbird)
*Chloropsis cochinchinensis jerdoni* (Blyth)
*Birds of Asia*, Vol. III, Parts XIII–XVIII, by John Gould, 1861–66. Painted by John Gould & Henry C. Richter.

In Memory of Nanubhai Bhailalbhai Amin, a true conservationist, from Bhailal Amin Foundation

20 minutes later when the herd took alarm and went off the calf galloped after its mother.

R.C. Morris

Honnametti Estate, Attikan P.O. Via Mysore, S. India,
5th January 1933.

[In an early issue of the Journal there is a reference by Mr. J.D. Inverarity to a new born bison calf which he came across. It lay crouched in the long grass with its neck stretched along the ground – a position in which it was not readily noticeable – its light yellow colouring blending with the dry grass. The mother being disturbed left it and joined the herd which Inverarity was tracking. The herd made a circuit of a mile or two and eventually came back to the place where the calf had been left. – Eds.]

*JBNHS*, Vol. XXXVI.

## Wild Dogs Killing a Panther

Yesterday I came across an interesting case of a panther having been killed by wild dogs. Apparently what occurred was this: A pack of wild dogs were lying up in some thick cover, and one of their number went down to water, and was immediately pounced upon by a panther. The noise it made when it was caught brought the whole pack down. In the furious fight that followed the panther was torn to pieces. I discovered no traces of any dogs having been killed in the fight.

R.C. Morris

Honnametti Estate, Attikan P.O. Via Mysore, S. India,
15th February 1933.
*JBNHS*, Vol. XXXVI.

**Dhole, Wild Dogs.** *Routledge's Picture Natural History*, by the Rev. J.G. Wood, engraved by the Dalziel brothers, 1885.

**Common Starling >**
*Sturnus vulgaris*

A glossy black bird, looking rather as if oiled, and more or less spotted with buff, giving the bird a pleasing speckled appearance. During winter the whole plumage is iridescent, with a high gloss of red, purple, green, blue. In summer the plumage looks blacker. Sexes alike but female is generally duller and more spotted. Gregarious, collecting in large flocks in winter, the starling is a bird of very wide distribution in Europe, Asia, and Africa. The Common Starling of Europe visits Sind, Gujarat, and Rajasthan during winter; in summer it visits Kashmir.

Common Starling *Sturnus vulgaris* Linnaeus
*Birds of Asia*, Vol. IV, Parts XIX–XXIV, by John Gould, 1867–72. Painted by John Gould & Henry C. Richter.

Courtesy Godrej Industries Limited

# The Ashoka Tree

The vernacular synonymy in the article on "Beautiful Indian Trees" (*JBNHS*, XXXVI, 2) gives *Ashopalava* as the popular Gujarati name of "Asoka" (*Saraca asoka*) tree. This, I am afraid, is not the case because the tree which is generally known as *Ashopalava* in Gujarat appears to be an altogether different species from the one illustrated in the Journal. The flowers of *Ashopalava* are creamy white – white with a yellowish tint. The leaves resemble those of a mango tree (vide specimen) and when young are a wonderful sight – the whole tree being enveloped in shining dull red foliage. The fruit (as you will see from specimen sent under separate cover) resemble unripe jujub (Guj. *bor*; *Jujuba jujube*). I think *Ashopalava* is a species of the Ashoka though different from the one described by the learned authors. Here I may add that ancient Indian writers on medicinal plants describe two species of Ashoka – one with brilliant red flowers (called "Hema Pushpa" Golden flowered) i.e. Ashoka (*Saraca asoka*) proper, and the other with white i.e. *Ashopalava*. [The specimen sent by the writer was identified as *Polyalthia longifolia* Benth. and Hooker f. – Eds.]

*Ashopalava* is also different from the Ashoka tree about which so much is written by Samskrta writers. If there is any tree which has received most attention in Samskrta love poetry, it is Ashoka. Its flowers are compared to the ruddy heels of a young and beautiful woman. A riot of colour, the very sight of these golden flowers is considered to increase passion. They form one of the five arrows of Puspadhanvan, the Indian God of Love (the remaining four are the Red Lotus, Mango, *Navamailika* and Blue Lotus). Indian Plant Lore also says that the flowering of an Ashoka tree is dependent on the gentle kicking of the tree by a young energetic and beautiful damsel whose ankles are ornamented with Nupuras. To continue, the other unfortunate (shall we call them fortunate) trees and plants which have to depend for their flowering on the sweet will of the fair sex are:–

*Tilaka* tree flowers when a beautiful woman glances lovingly at it.
*Kuravaka* (a species of Amaranth), when embraced by a lovely maiden.
*Priyangu*, by the gentle touch.
*Bakula* (*Mimusops elengi*), when sprinkled with wine from the mouth.
*Mandara* (Coral tree *Erythrina indica*), by witty remarks or light pleasantry.
*Champaka*, when a lovely woman gently laughs near it.
Mango tree, when fanned by the gentle fragrant breath of a beautiful damsel.
*Nameru* (Rudraksha tree), hearing sweet music.
*Karnikara*, when a damsel dances in its presence.

(By the by I may mention that this strange belief in flowering of certain plants through human agency has been explained away by a Samskrta writer as the peculiar experiments conducted by experts to make plants, trees, etc., flower before their time.)

The Ashoka tree also figures in religious rites. Hindus, particularly women, observe a vow called "Tri Ratra" (for three nights) for the removal of misery and impending danger. "Ashoka Purnima" or 15th of the bright half of the month of Phalguna (March) is celebrated in honour of the tree, as an aid to love. The 6th day of the bright (also dark) half of Chaitra (April) is observed to get happiness and the blessing of a son. The 8th of the bright half of Chaitra is also observed in honour of Ashoka.

Hari Narayan Acharya

Ahmedabad, 24th May 1933.
*JBNHS*, Vol. XXXVI.

## A Leopard-like Tigress

The following account of a "leopard" on a neighbouring tea garden may interest your readers. The animal first appeared about Christmas time, and for two months stayed in the vicinity, killing at least five calves and one pig, all in houses. On one occasion after killing a calf in a house and taking it out, the marauder was chased away, but returned at 2 a.m. and was fired at and missed. Yet it returned again to take away the kill before daylight.

During a space of ten days this animal appeared four nights in the manager's bungalow compound, and endeavoured to break into the hen-house. Once it appeared at the steps of the back verandah. One night while in the compound two shots were fired at it, both missed.

Reports of those who had seen the animal agreed that it was a leopard with a tiger's face. Thereupon my friend devised a portable all-iron goat trap to catch this extraordinary creature. Almost immediately the animal was caught on March 9th and turned out to be a tigress, 8 ft. 3 in., nearly dry of milk. Cubs were supposed to have been heard of in the district.

She was in an emaciated condition, and covered with ticks, some the size of a 3-penny piece. Is it not unusual for a tigress, even though starving and with cubs, to take to the habits of a house-invading leopard? Presuming she had not the strength to kill a buffalo or large bullock, there were thousands of cows and calves throughout the whole district on the fringes of the jungle – an easy prey to the cattle-lifting tiger.

E.P. GEE

Badlipar P.O., Assam, April 11, 1937.
*JBNHS*, Vol. XXXIX.

## Smoking a Panther to Death

I was staying at the hill fort of Sinhgar near Poona for the summer. There are in the jungle near about small panthers and other wild animals. One day, when we were out shooting, a small panther was observed in the beat and was tracked by its pugmarks to a small cave. In order to dislodge him, we smoked the cave at one end, the guns being posted at the other. As the panther did not come out, it was thought that the animal had lodged itself near the opposite opening of the tunnel, which also was smoked. After a little while a gasping sound was heard from within. The expert tracker with us said that it was the death gasp of the panther. We wanted to go in to see what had happened, but the son of the shikari, who was watching the proceedings from the top of a tree became suddenly possessed by the local deity and ordered the people to desist from entering the cave until the next morning. It was sound practical advice which we obeyed. No watch was kept at the cave. We did not believe the story of the shikari, but next day, to our surprise, the villagers, in the company of one of the guns, went into the cave and dragged out the carcase which they brought to the fort where I was staying then. If I remember rightly the above incident was reported in the local Poona English papers of the day.

CHIEF OF ICHALKARANJI

Camp Madhavgiri, Ajra, 11th May 1937.
*JBNHS*, Vol. XXXIX.

## The Mating of Elephants

While out shooting wild elephants in 1919 on the west of Payagale, about 20 miles north-west of Pegu, Burma, I chanced to witness a unique sight, and was sorry to be without a camera at the time. It was the mating of a wild bull-elephant and cow-elephant. I was searching for a tusker that morning and came upon a herd of wild elephants which were scattered grazing. Entering stealthily in their midst followed by two Burman guides, I saw from a hill streamlet, after an hour's stalking, two wild elephants – a bull tusker and a cow – far removed from the herd just at the foot of a hillock. The bull had its trunk round the left hind leg of the cow-elephant, at the same time pressing his right tusk on her left rump, using this as a lever to inflict pain so that the cow would be obliged to submit to his wishes. No resistance was shown, apparently because the bull had a thorough hold on her.

From my place of vantage, say within 50 feet away, I saw the bull forcing the cow, held in the above manner, to walk up the rising ground for about the space of 20 or 30 yards; they turned and the bull loosening his hold, rose on the cow in the act of service. They gradually descended in this manner to the foot of the hill. On reaching level ground the bull got off, laid hold of the cow again as described above and made her repeat the ascent. This movement was carried out three or four times in succession till the service apparently came to an end.

The wild bull then released the cow which ran away to join the herd once more. The bull stood for some time before he moved off.

The scene was so impressive and unique that I was loath to shoot the tusker, and allowed him to go his way. I don't think such a sight has been witnessed by many shikaries, and I therefore record this note.

J. Gonzalez

No. 10, 3rd Street, Pegu, 27th November 1938.
*JBNHS*, Vol. XL.

**Mottled Wood Owl >**
*Strix ocellata*

A large owl with a typical "owl face" but without ear-tufts; beautifully mottled and vemiculated above with reddish brown, black, white, and buff. Facial disc white, finely barred concentrically with black; ruff white and black with chocolate admixed. Below, throat chestnut and black stippled with white; prominent white half-collar on foreneck. Rest of under-parts white and golden buff with narrow blackish bars. Sexes alike. Nocturnal. Pairs spend the day sitting together and dozing on a secluded branch hidden in the foliage. When disturbed will fly a long distance in bright sunlight without any apparent discomfort. It lives entirely on squirrels, rats and mice, and medium-sized birds. Lives in open wooded areas and groves of large trees around villages and cultivation. The call is said to be a loud harsh hoot. Resident, endemic to the subcontinent. From E. Rajasthan east to Bihar and south through the peninsula, except in E. India; also locally in N. India and S. Gujarat.

## The Hatching of a Mugger (*Crocodylus palustris*)

In May this year we were camped at Devikop forest bungalow about 20 miles from Hubli on the Hubli-Yellapur Road (Kanara). Below the bungalow there was quite a large jhil which does not dry up during the summer. Between the bungalow and the water and about 70 yds. from the water there was a small experimental teak plantation. The Forest Department were working in this plantation putting in teak seeds. One of the labourers engaged on this work was digging when he came across a nest of about 20 crocodile eggs. We opened one and found a fully formed young crocodile which would probably have been born in a few days. As a matter of interest I kept one. On return to Belgaum I put this egg into a cup on a shelf in my sitting room and forgot about it. I had been back here three weeks and was having my breakfast one morning when my bearer came and said there were noises coming from the egg. This was quite correct, and I realized there was a live crocodile in the egg. I then put the egg in a biscuit tin and placed it in a warm place; the noises continued for several days and a very small crack appeared in the shell. When I saw this I took a knife and removed some of the shell. I then put the egg back into the tin and the next morning had a look but it was still in the same condition. I had another look an hour later

Mottled Wood Owl *Strix ocellata* (Lesson)
*Birds of Asia*, Vol. IV, Parts XIX–XXV, by John Gould, 1867–72. Painted by John Gould & Henry C. Richter.

In Memory of Bonny Shah, who loved birds, animals, trees and plants, and also children,
from Mr. Ratilal Shah, Ahimsa of Texas, Bartonville, Texas, USA

and the crocodile was born fully formed and full of life. He is now in his natural surroundings.

C. BONE, CONDUCTOR

Officers' Training School, Belgaum, 8th June 1943.
*JBNHS*, Vol. XLIV.

# The "Courtship" of the Monitor Lizard (*Varanus bengalensis*)

The accompanying photographs were taken by me at Chaduva in Kutch on August 17, 1943. The monitor lizards concerned were rather over 2 ft. long each. One of them, with the tail-tip missing, was the heftier of the two and the more aggressive. Him I assumed to be the male, the other the female. Ostensibly the lizards were engaged in an "all-in" wrestling match, and many of their grips, catches and throws were surprisingly human. The commonest manoeuvre was to stand up on their hindlegs, clasping each other firmly about the neck and shoulders, and then with a sharp sideways jerk of the head to knock the other down – sometimes tossing it completely over. The victor, who invariably happened to be the "male", now appeared to try and twist the posterior end of the "female" round into a position suitable for copulation. The struggle, which was interspersed with much bloodless biting on the neck behind the ear, lasted without result for over an hour and a half. Both combatants were panting heavily and were visibly exhausted. Occasional pauses occurred only when the female – who seemed to be more timid than the other – walked away upon my approaching closer with the camera. The male seemed unperturbed by my proximity, at one time under 4 feet. On these occasions the male did not follow or attempt to chase her but stalked slowly over the "ring" nose to ground, body raised to full height, as if smelling. After retreating a few yards, and within the space of 2 or 3 minutes the female, though she had appeared to be having the worst of the encounters all along, returned to "Hefty" and the bouts recommenced. This circumstance is enough to suggest that it could not have been a serious fight but some sort of rough courtship that was in progress. After the ungentle handling the female had received it is hard to imagine her returning to the fray of her own accord when she had such a glorious opportunity for escaping. Although "Hefty" had apparently been winning the whole time he was the first to show signs of exhaustion and soon afterwards was completely done up. In the final stages the female took hold of his foreleg in her jaws and shook it violently from side to side two or three times, just as a terrier shakes a rat. Upon his still continuing inert she left him with what seemed a frustrated and disgusted

### Grass Owl >
*Tyto capensis*

A grassland owl, resembling a Barn Owl in size and structure, with dark eyes and whitish facial disc bordered by a dark brown ruff. Above, dark brown minutely spotted with white; below, white with scattered brown spots. Tail largely white and buff, cross-barred with brown. Legs very long and tightly feathered and in the words of Dr Sálim Ali, "as if clad in underpants or churidar paijamas!" In flight it looks white with brownish patches. Sexes alike. Crepuscular and nocturnal. Spends the daytime standing bolt upright and dozing amidst tall grass, flying a short distance when disturbed and dropping into the cover again. Food: chiefly field mice, locusts, grasshoppers and cicadas, and also small birds. Resident, from Uttar Pradesh east to Manipur; E. and SW India.

Grass Owl *Tyto capensis* (A. Smith)
*Birds of Asia*, Vol. IV, Parts XIX–XXIV, by John Gould, 1867–72. Painted by John Gould & Henry C. Richter.

Courtesy London Star Diamond Company (India) Pvt. Ltd.

sneer, eloquent of much damaging reflection upon his virility! She then deliberately waddled off up a sloping bank and into the shrubbery. It took some minutes for "Hefty" to recover himself when he too walked away dejectedly in a different direction and with no attempt to follow her.

It would be interesting to learn from someone who has studied the habits of Monitor Lizards whether this was in fact some courtship proceeding and whether such ordinarily leads up to mating.

SÁLIM ALI

33 Pali Hill, Bandra, 7th December 1943.
*JBNHS*, Vol. XLIV.

# Rarity of Man-eating Tigers in South India

Why are Man-eaters so rare in South India?

There have been, and may still be, many man-eating tigers in the Ganjam District, of course, and part of this district is south of a line drawn due east from Bombay. Vizagapatam District has had its man-eaters, and also other parts of the "Agency Tracts".

The Nallamallais, Kurnool District, provided the man-eating tigress shot eventually at Diguvametta by the then Conservator of Forests in September 1923. She preyed upon the luckless railway gangmen; and this habit brought about her death. The Conservator was told to walk along the railway embankment keeping a sharp lookout on both sides. He soon spotted the tigress making for a culvert ahead of him. He walked towards the culvert, and then over to the other side of the embankment: and stopped. The tigress, misjudging his position, popped up ahead; and received her quietus. She was in good condition, but carried an old scar.

A man-eater roamed the Baragur Hills, to the east of the Biligirirangans (Coimbatore District), some 30 to 35 years ago; killing people spasmodically – about 4 or 5 a year. A Government Notification offering a reward for the brute described it as "Ashy-grey, and somewhat stout"! It was said to have been shot by a poacher: if so it was quickly succeeded by another man-eater (not an unusual case) which also killed humans at infrequent intervals from Talamalai north-eastwards to Madeswarammalai and Ponnachi (Kollegal Taluk, Coimbatore District). Lt.-Col. R.E. Wright and I went after this tiger, reputed to have a kink in its tail. One night our camp was pitched in a field at the western foot of the Baragurs, an ill-chosen spot infested with masses of hairy-caterpillars. We went to sleep in two small tents facing each other, with a "Petromax" lantern burning between the tents; and loaded rifles by our cots. I was awakened at midnight by a horrified yell from R.E.W. and rushing out, collided with him; on which he collapsed with laughter. Half-asleep he had imagined seeing a large form stealthily moving into his tent; it turned out to be the shadow cast by the petrol lamp on the wall of the tent, of a large caterpillar crawling over his bed clothes!

Now this tiger *was* shot by a poacher. Sallying forth after deer in the early morning he met the tiger round a corner, fired his muzzle loader at it, dropped the gun and fled like the wind. Later in the morning a Forest Guard and his watcher, on beat duty, came on the dead tiger, and recognized it as the man-eater. The gun was also recognized by the watcher who named its owner. So a bargain was struck. The poacher was told that he would not be reported for being in the Reserve Forest with an

**Spotted Nutcracker >**
*Nucifraga caryocatactes*
A crow-like, dark chocolate-brown bird, streaked and spotted with white. Wings black, the white of the outer tail feathers being conspicuous in flight. Stout, wedge-shaped bill. Sexes alike. Patchily distributed between 2,000 and 6,000 m; moist temperate and alpine conifer forests of pine, fir, and spruce. They remain in pairs or family parties of 4 or 5 in tall conifers on hillsides. Food: seeds of conifers, nuts, eggs and nesting birds, etc. Shy and wary, the loud, grating call attracts notice a long way off. Resident, mountains of NW Pakistan, and Himalayas from N. Pakistan to Arunachal Pradesh.

Spotted Nutcracker Nucifraga caryocatactes (Linnaeus)
*Birds of Asia*, Vol. IV, Parts XIX–XXIV, by John Gould, 1867–72. Painted by John Gould & Henry C. Richter.

Courtesy ITC Limited

**Rosy Minivet** *Pericrocotus roseus* (Vieillot)
*Birds of Asia*, Vol. II, Parts VII–XII, by John Gould, 1855–60. Painted by John Gould & Henry C. Richter.

Courtesy Suresh Pethani, M. Suresh & Co.

unlicensed gun, provided the Forest Guard was given the skin and skull of the tiger. This was gladly agreed to by the poacher who had no idea that the tiger was a notified man-eater, with a reward of Rs. 300 on its head. The F.G. then proceeded to claim the reward, producing the skin and skull in the local Katchery together with a wonderful story of how he had killed the tiger single handed sitting up for it over a jungle path. The reward was about to be paid when the watcher learned that the F.G. intended to give him a mere pittance of the total. So the watcher then "blew the gaff". The only party to benefit was an unkind Government who, while pardoning the poacher for his activities, confiscated his gun and paid out no reward – the F.G. being sacked, and the tiger's skin and skull retained in the Katchery.

In more recent times, two years ago in fact, a tiger killed four or five people at the western foot of the Billigirirangans (Mysore District) and was finally shot by officials. Earlier this year a tiger killed three persons in the Talavadi firka, Gobichettypalayam Taluk, and was finally shot in Mysore territory.

The foregoing still does not explain why there are fewer man-eaters in South India – but I think the answer possibly lies in a combination of circumstances. Continued existence of man-eaters in an area where both game animals and cattle exist in insufficient numbers, and where tiger are forced to remain instead of emigrating to more fruitful parts – as in the case of Ganjam for example. Sanderson's description of the Honganur (Mysore District) man-eating tigress (in his "Thirteen Years among the Wild Beasts of India"), which he finally shot, is an example of the vice picked up by a tigress to provide easy food for her cubs in all probability.

R.C. MORRIS

Honnametti Estate, Attikan P.O. Via Mysore, S. India,
13th December 1945.
*JBNHS*, Vol. XLVI.

< **Rosy Minivet**
*Pericrocotus roseus*

Male rosy pink with ashy brown upper-parts. Female like male but rosy parts replaced by pale yellow. Arboreal and more sluggish than other minivets. Affects deciduous or evergreen forests, lightly wooded country, and gardens. Call: a squeaky whistling *whiriri-whiriri-whiriri* (Sálim Ali and S. Dillon Ripley). Occurs from western Himalayas to Arunachal Pradesh and other parts of NE India; winters in the peninsula.

# Fishing with the Indian Darter (*Anhinga melanogaster*) in Assam

In February 1948 I met a party of about twelve men journeying along the main Assam trunk road, and carrying with them six Darters (*Anhinga melanogaster*). The birds were completely tame, and were quite oblivious of passing traffic. They were carried on long bamboo yokes across their owner's shoulders, as the accompanying photograph shows.

I was informed that the party came from the district of Dhubri in north-west Assam, where they use their Darters as a regular means of catching fish in the swamps and small lakes adjacent to the Brahmaputra. Since they were some sixty miles from their village when I met them I enquired why they were travelling with their birds. They told me that during the dry season they often leave home and "vagabond" through the country, catching fish as they go and trading it in on the spot for rice and other goods. The party was quite self-contained when I met them, and those who were not carrying birds had bedding, cooking utensils etc. I have since been informed by Mr. W. Shaw M.B.E., A.C.S. (Retd.) that when he was district officer at Dhubri some years ago he often saw the local villagers using their Darters; first putting a ring round their neck, and then launching the bird. On returning with a fish it was at once given a small piece before sending it out again.

The men I met were Hindus, but Mr. Shaw tells me that he has most often seen Muslims fishing in this way; so that there seems to be no question of a special "Darter-fishing" caste.

Two reliable informants have also told me that either Darters or Cormorants are used in Manipur State on the Burma border for catching fish in the great Logtak Lake.

C.R. STONOR

7th March 1948.

[Fishing with trained cormorants is still commonly practised in China, and to a limited extent also in Japan. We were not aware of Darters being employed for the purpose or that this "biological" method of fishing was in vogue anywhere within our limits. Why Darters should here be used in preference to cormorants is not clear, since the former's narrow head and slender neck naturally makes it possible for it to tackle only much smaller fish than the cormorant can. – EDS.]

*JBNHS*, Vol. XLVII.

# "Pandadi" or *Strobilanthes callosus* (Nees) at Junagadh in Saurashtra

Generally *Strobilanthes* species grows in localities having high altitudes. In Saurashtra there are several hills of volcanic origin of trap rock formation, but of all these hills, *Strobilanthes callosus* Nees grows exclusively on the slopes of the valley formed by the sacred hills of the Girnar and Datar ranges. This plant is not seen growing on the top of these hills. This interesting economic plant which occurs at Junagadh is not recorded by Cooke, 1905, but Thakar, 1926, has made a passing remark about it. According to Sutaria, 1949, and Vaidh, 1945, it occurs in Gujarat.

Locally the plant is known as Pandadi. As it occurs on Girnar hill ranges it is cited as Girnari Pandadi by Vaidh, 1945, and Kirtikar *et al.*, 1933 and as Junagadhi Pandadi by Thakar, 1926. So its Gujrati name is "Pandadi" and not "Karvi" as cited by Sutaria, 1949.

Once upon a time Junagadh was famous for its Pandadi oil and it had attracted experts from Kanoj, who used to come here every year for some time for the extraction of Pandadi oil. This plant has got medicinal value also. Local vaids do not make use of the plant but hakims use it as one of the ingredients in the preparation of an ointment for boils. It is used by the general public for its aromatic and insecticidal properties. But no mention is made about this plant by Chopra *et al.*, 1941. It is used to protect woollen fabrics from insects. It may be an insect-repellent rather than an insect destroyer.

For noting the flowering cycle, a shrub of *Strobilanthes callosus* Nees was brought from the hills and planted in the College garden in the monsoon of the year 1943. This plant flowered for the first time in the month of August 1949, i.e. after six years. The flowering season lasted up to the month of November. Then the plant entered into fruiting stage and by the end of February 1950 all the leaves disappeared leaving only dried fruits in the axil of the persistent bracts. When young these bracts were green and glabrous; but when they grew old they became brown and were seen densely covered with glandular hairs. Tips of these hairs, were seen bedecked with transparent

**Karvi** *Carvia callosa*
A small rugged shrub, Karvi mass-flowers in the eighth year of its existence. Whole hillsides from about Baroda in Gujarat south to about the Nilgiris burst into bloom in August. Thereafter, having broadcast its seed, the plants die out. Karvi serves well as an indicator of true monsoon rains as no amount of pre-monsoon showers can stimulate it to regenerate, but with the commencement of the true monsoon the rootstocks sprout. Each mass flowering is preceded by an odd plant flowering here and there in the year before. (Information from the late J.S. Serrao.) Photograph by Ashok. S. Kothari.

globules of a viscous substance which sent out strong smell of balsamic nature. Local belief here is that only those bracts which come in contact with dew, produce an odour.

G.A. KAPADIA

Bahauddin College, Junagadh, 20th March 1950.
*JBNHS*, Vol. XLIX.

## The Flowering of *Strobilanthes*

In a previous note in this Journal I reported a general flowering of *Strobilanthes callosus* Nees that had taken place in Khandala during the summer months of 1943. At present I can add some more details that may be of interest to the members of our Society.

Flowering did not seem to depend on altitude, or exposed situation of the place, or even size or age of the plants. The whole phenomenon may be summarized in the following points: 1. In 1942 a few "precursors" or "forerunners" came into flower in various parts of Khandala, a few plants at a time, as Mr. C. McCann told me at the time, "announcing a general flowering in the near future." 2. A general flowering took place in most parts of the district in 1943. 3. An almost equal general flowering again took place the following year, 1944, in places where plants had not flowered the previous year. 4. The flowering cycle seemed to close with a few stragglers that bloomed in 1945, a few plants at a time, scattered throughout the district.

To my great surprise and pleasure, in summer 1949 I found in Khandala two clumps in flower, the first below Elphinstone Point, and consisting of only 40–50 plants; the second was on top of Bhoma Hill. Recently I examined the second clump consisting of many thousands of plants, and found them in fruit.

The top of Bhoma Hill is roughly a triangle with fairly broad or obtuse corners; the edges and slopes of this triangle are covered with almost pure stands of *S. callosus* Nees (*Carvia callosa* Bremek.), the centre of the hill-top is but a grassy plateau. One side of the triangle goes almost perfectly E to W, the second side W to SE, the third side E to SW. Approach to the top of the hill is by a path that passes through Forebay and the "Saddle" and enters the plateau by the western obtuse corner; the patch then continues W to E parallel to the north side of the triangle and descends by the eastern corner, Barometer Hill.

In last year's flowering it was noticed that only the plants in the shaded portion of the diagram had flowered; the dividing lines going W–E and N–S are imaginary, nevertheless they are both astonishingly clear: all the plants west of the NS and south of the WE lines showed masses of fruits, whilst all the neighbouring plants E or N of the line were bare of fruits and had obviously not flowered last season; this is remarkable because even when plants were touching each other across the imaginary line they showed such an independent behaviour. Climatic or edaphic conditions seem to be exactly alike on both sides of the line, yet the clearly different behaviour of plants demands an explanation, which for the present I am unable to give.

Another point of interest is the length of the period intervening between two flowering seasons. The plants that had bloomed in 1949 were certainly in flower in 1943 or 1944; this reduces the flowering period to six or seven years. This coming summer and the following I shall try, and keep careful watch for the possible general flowering and the exact spots where it may take place.

In 1943 and 1944 I received reliable reports or personally noticed signs of a general

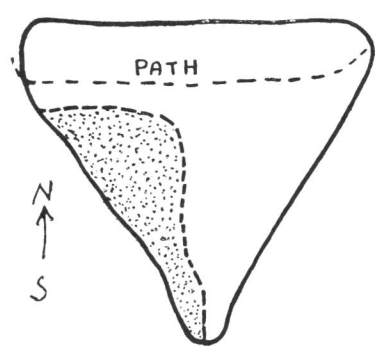

Diagram showing the top of **Bhoma Hill**.

flowering at Mt. Abu, Purandhar, Khandala, Matheran, Kanheri Caves, Castle Rock, and as far south as the Nilgiris. May I request readers in various parts of India, where this species is common, to keep a watch over these plants and report any general flowering? It is only through such concerted action that this intriguing problem may finally be solved.

H. SANTAPAU

St. Xavier's College, Fort, Bombay, 31st March 1950.
*JBNHS*, Vol. XLIX.

# The "Dew-Claws" of the Hunting Leopard or Cheetah

I have not had the fortune to witness the sport of blackbuck hunting with aid of the cheetah. None of the accounts contained in shikar books are available to me just now so I turn to the "Fauna of British India, Mammalia", Vol. I where the author, R.I. Pocock, F.R.S., describing the method of hunting from accounts available to him (for he will not have written from personal observation of the sport) says at page 329:

"The victim ... is usually apparently struck over by a blow of the Cheetah's fore-paw, is then seized by the throat ...."

One would think that the forearm and paw of the cheetah has not the muscular power necessary for such a feat, especially when it is borne in mind that the weight of the buck is about 90 lbs. It seems that it has not been related in the accounts available to Pocock exactly how the buck is struck down in full flight.

I have just come across the two volumes of a book titled "The New Shikari at Our Indian Stations" by Colonel Julius Barras, 1885, and read at page 92 of Vol. I exactly why the cheetah is able to strike down the buck. Having described the preliminaries and circumstances of the hunt Barras relates:

"I now inspected the carcass of the deer (*sic*) with a view to ascertain if possible how the cheetah had been able so instantaneously to strike down such a powerful animal immediately on getting up with it. I at once observed a single, long deep gash in the flank which was evidently caused by the decisive blow. But I could not imagine with what weapon the leopard had been able to inflict this very strange-looking wound. Then, turning to the beast, as it sat on the cart, I inspected it closely and saw that the dew-claw which in the dog appears such a useless appendage, is represented in this brute by a terrible-looking talon exactly suited to the infliction of such a gash."

So here we have it. It is not, as we can readily imagine, by a blow of the paw alone that the buck is struck down: Nature, to aid the cheetah's speed has provided almost dog-like nails to his four toes, but has retained for him the powerful, sharp, curved dew-claw to enable him to obtain the necessary purchase to overthrow the buck at racing speed. Without such a dew-claw the beast would probably not be able to strike the blackbuck down with the sureness he displays.

R.W. BURTON, LIEUT.-COL. I.A. (RETD.)

Bangalore, 10th July 1950.

[K.S. Dharmakumarsinhji of Bhavnagar who has considerable experience of hunting with trained cheetahs, comments as follows:

"The cheetah's main weapon of attack is the dew-claws without which it would be difficult for him to hold down large prey.

"Our experience in hunting with cheetahs is that the dew-claws are made full use of as hooks for holding on to blackbuck once the animal has been contacted. Cheetahs with blunted dew-claws were not able to control full sized blackbuck as effectively as those that possessed sharp undamaged ones. We have found therefore that the dew-claw is very important to the cheetah and he can also inflict a severe wound with it.

"Sometimes it is solely by means of the dew-claws that the blackbuck is secured in the chase."
– Eds.]

*JBNHS*, Vol. XLIX.

## Rabies in Tiger – Two Proved Instances

In all the years of the Bombay Natural History Society since its foundation on the 16th September 1883, and the issue of the first Journal in January 1886, there has been no instance recorded of rabies in either tiger or panther. Now we have from Assam two proved cases of rabies in tiger, and a possible third instance. (Calcutta *Statesman* of 9th February 1946 and Miscellaneous Note by Mr. S.A. Christopher in Vol. 46, p. 391.)

There has also, so far as known to me, been no mention of rabies in tiger or panther in any of the more than 250 books published during the past 150 years on "Big Game Hunting and Shooting in India and the East." (Vol. 49, pp. 222–240.)

This is very strange, for it can be reasonably conjectured that cases must have occurred. Also one would think that panthers, being so partial to the killing of domestic dogs for food, would have been very many times exposed to the possibility of contracting rabies through the fresh saliva of their prey. Possibly the known instances of tigers and panthers being found in the forests, and cause of death not apparent, were due to rabies.

*First Case.–* The narrative (condensed) of Mr. T.R. Clark, the then Manager of the Salonah Tea Estate in the Nowgong District of Assam, gives particulars of what happened.

"On the evening of 28 January 1943 one of three men cycling in the dark along the Salbari Road was attacked by a tigress and badly mauled. The two other men managed to drive off the animal, the cycle held up by one of them as a shield being bitten and clawed.

"Further up the road some carts, carrying long mats extending well over the cart buffaloes, were attacked by the tigress which leapt from the bank at side of the road on top of the mats and was perched there for several yards. The men, sheltered by the long mats, managed to scare the animal away. Shortly after, more carts carrying smaller mats came along. The tigress leapt on the buffaloes, dragged down one of the cartmen and badly mauled him. In the terrible confusion three of the men were seriously clawed and bitten, one of them suffering a fractured arm.

"The cyclist was admitted to hospital early in the night and the cartmen were brought in about 8.30 next morning (29th Jany.) Not long after I motored along the Salbari road to see the places of attacks, and going on towards Langteng met Mr. Edwards, Assistant Manager, bringing in a woman and girl badly mauled on the road about 8.45 that morning. These having been admitted to hospital I sent Mr. Edwards to Nowgong, over twenty miles distant, with a letter to the Deputy Commissioner asking for elephants and shikaris.

"About 2 p.m. a man cutting wood was brought in badly mauled. Later in the afternoon the Deputy Commissioner arrived, and Dr. Hugh Smith (Medical Officer, Nowgong Medical Association, Salonah) and I went with him to see the place where, it was just then reported, another woodcutter had been attacked. This was several miles distant from the previous place.

"Arrived near the place it was getting dark so a lorry and other three men were obtained. With much difficulty the lorry was forced through as far as a small clearing, and after searching around for some time in the dark the man was heard to be faintly calling in reply. He was found to be badly torn, but conscious and sensible. The tigress could not have been far off, and having but a torch to show the way through rough scrub and jungle the party was rather helpless had the animal attacked, for firearms in such circumstances would have been difficult to use with proper effect.

"At 6.15 next morning (30th) I went to Borghat to take two ladies to the railway station for the Darjeeling train. Passing through the estate labour lines it was found there was much excitement as the tigress had been seen in the vicinity.

"Returning through the line with the ladies I stopped for information. Suddenly the chowkidar pointed, and there was the tigress quietly crossing the road about ten yards away and not even glancing towards the car! I rushed the ladies to the station, put them safely into the stationmaster's office, and then started off to get my gun from my bungalow and Dr. Smith with his heavy rifle. On my way I met Mr. Rogers of Amluckie Tea Estate who carried on to the station with his wife, who was also going to Darjeeling, while I proceeded to my bungalow. Mr. Rogers fortunately had his gun with him.

"Now came the almost incredible climax. Arrived at the station Rogers got the luggage out of the Ford Vanette, turned the car round, and was sitting in the driving seat, his gun unloaded, when the tigress came walking along the road towards the car. He was unable to load the weapon in the confined space and could not risk getting out. The beast walked past, taking no notice of the car or of two bullocks tethered to a close-by cart. Rogers got out of the car, loaded his gun and shot the tigress as it walked slowly and quietly away from him.

"We learned later that the tigress had forced her way into houses in the lines; and in one instance, where a man with his wife and child tried to close the door against it, mauled the woman and child but did not attack the man.

"The tigress was fully grown, perhaps three years old. The brain was quickly removed and sent by train, held up for the purpose, to the Pasteur Institute at Shillong. Medical Officer (Dr. Smith) reported "*Negri* bodies in tiger's brain" thus furnishing proof positive that the animal was suffering from rabies. (Note: *Negri* from "Negroid" – black bodies seen under the microscope.)

"All the injured people were given anti-rabic treatment. Fifteen cases were treated in the hospital and eight of them died of their wounds. Two Mikirs (people of the Mikir Hills) and a Nepali woman were killed in the jungle. None of the survivors developed rabies. During the thirty-six hours of the "terror" eighteen people were attacked of whom only seven survived. The cyclist, the first man to be attacked, died at the very moment the sound of the shot which killed the tigress was heard at the hospital. This coincidence quickly gave rise to much talk and conjecture among the superstitious!"

**Short-billed Minivet >**
*Pericrocotus brevirotris*

Male black and scarlet with black throat, a broad scarlet band running through the black wing, a black and scarlet tail and scarlet lower back. Female has a grey back, yellow throat, and all the red parts of male replaced by yellow. Purely arboreal; found in flocks which attract attention by the bright plumage. Distinguished from the Scarlet Minivet by the smaller size, slimmer build, greater amount of black in the tail, and absence of scarlet (in female, yellow) spots on the wings. A shrill persistent *tiwiwiwi* given by both sexes when perched or in flight. Resident, Himalayas from Uttaranchal to Arunachal Pradesh, Mizoram, Nagaland, and Bangladesh; c. 1,800–2,400 m, descending to foothills in winter.

**Short-billed Minivet** *Pericrocotus brevirotris* (Vigors)
*A Century of Birds from the Himalaya Mountains*, by John Gould, 1832. Painted by Elizabeth Gould.

In Memory of Vijaya Deshmukh, a true lover of nature, from Mr. B.G. Deshmukh, President, BNHS

White-bellied Minivet (Cawnpore Minivet) *Pericrocotus erythropygius* (Jerdon)
*Birds of Asia*, Vol. I, Parts I–VI, by John Gould, 1850–54. Painted by John Gould & Henry C. Richter.

Courtesy Poultry Development Promotion Council

*Second Case.–* In a letter published in the Calcutta *Statesman* of 6th May 1950, Mr. M.N.R. Kemp of Saikhoa Ghat, Assam, stated that a tiger had attacked a village during the night of 16th April 1950, and in a further letter published on the 20th May followed up with information that the brain of the tiger was reported by the Pasteur Institute, Shillong, to contain *Negri bodies*. It appears that the tiger went from house to house in the village, attacking, mauling, biting and going on to the next house. In all 14 persons were mauled of whom one, a woman, died at once and two others in hospital. One man was bitten in the upper arm and suffered a compound fracture of the humerus. His wife it was who was killed outright and the two elder children bitten, while a small baby was unhurt. The tiger was shot at dawn by a local shikari – the length 9 feet 4 inches.

**Pasteur Institute Report**

In his letter of 8th July 1950 to Mr. Clark the Director of the Pasteur Institute, Shillong, Dr. S.R. Pandit mentions that he quite well remembers the above two instances of rabies in the tiger as he had examined and reported on the sections from both. He mentions also that he has submitted a note of these two instances of proved rabies in the tiger to the *Indian Medical Gazette*.

*A Third Case.–* The Miscellaneous Note by Mr. S.A. Christopher under the caption "A Tiger 'Runs Amok'" draws attention to an article relating that a tiger killed 7 people and was then killed in battle with a wild buffalo. Such an occurrence – seven men all killed at one spot in presence of other people, presumably in daylight in or near a village and a railway station, is certainly deserving of further enquiry. On the face of it this seems quite likely to have been another case of rabies in the tiger.

**How Did the Tigers Contract Rabies?**

It is natural that people should wonder in what manner these tigers contracted rabies. As rabies virus cannot gain entry into the body other than through broken skin, and the virus has to be conveyed by means of *fresh* saliva, it follows that the two animals contracted the disease either by being bitten or wounded by some rabid animal; or (which is more likely) the virus in fresh saliva from some rabid beast they had killed, or fed upon, entered their bodies through some wound or abrasion, or some break in the skin of mouth or tongue. They may have licked the "kill" and so licked fresh saliva and in that way have got the virus into their system. The smallest break in the skin would suffice, and that may have come about in a number of ways: a cut from a sharp bone on some part of lips or tongue; a scratch on the lips from a thorn; or any of many possible happenings.

R.W. Burton Lt.-Col. i.a. (Retd.)

Bangalore, 16th September 1950.
*JBNHS*, Vol. XLIX.

---

**< White-bellied Minivet**
*Pericrocotus erythropygius*
Sparrow-sized bird with a long tail, found practically throughout India except extreme NW. The male is glossy black and white with a red patch on lower back. Below, beautiful rosy patch on the breast; rest of the under-parts white. In the female the black parts are replaced by dark brownish grey; forehead and lower plumage white; rump white and orange. Affects semi-desert and open grassy forests in arid areas. Less arboreal than other minivets. Keeps in small parties of 6 to 8. Call: a whistle like the *tseep-tseep* of wagtails; also a short sweet song (Dharmakumarsinhji). Resident, mainly Gujarat and from central Rajasthan east to Bihar and south to Maharashtra and Andhra Pradesh.

---

# Experiments in Implanting African Lions into Madhya Bharat

The lion in India used to be fairly common in the jungles now included in Rajasthan and Madhya Bharat. It is unfortunate that it is not found any longer in the country except in the Gir forest in Saurashtra. The reason for its disappearance is the tiger which kept on increasing in number and killed off or drove away the lion until it found an asylum in the Gir forest. This forest is an isolated area completely cut off by over a hundred miles from the tiger infested hills. This tiger is the kind of animal which does not allow other large carnivora feeding upon the same food to live in the same locality. It is like the case of having two swords in one scabbard.

The tiger seems to have come to India from China, Assam, Burma etc., through Bengal, and that is the reason why it is still called Bengal tiger. It was more cunning and powerful than the lion and therefore it killed off or drove the lion away from the areas it occupied.

I had a few opportunities to arrange duels between the lion and the tiger in a small arena specially prepared for the purpose. In three such experiments on three different occasions I found the same result. It is the lion that always makes the first attack and it is he who gets the worst of it. One or two smacks from the tiger are enough to make the lion retire.

The late Maharaja Sir Madho Rao Scindhia, realizing that lions had existed in his State (Gwalior) in the olden days, resolved to re-introduce them. With this object he imported three pairs of lions from Africa. The jungle selected was Sheopur and Shivpuri forest range, which covered an area of some 1,000 square miles.

When these animals arrived they were taken to a place called Dobe Kund which is practically half way between Sheopur and Shivpuri. A special enclosure of stone wall, 20 ft. high, was prepared, in which the lions were kept. They were not fed on dead meat but were always provided with live buffaloes so that they might not lose the natural habit of killing animals. They were kept in this enclosure for about 4 years during which they not only got thoroughly acclimatized, but also bred and increased in number.

This place was situated in a lonely spot in the midst of forest abounding in tigers. The roaring of the lions always attracted the wild tigers, but on account of the high wall they could not get at them. We used to make periodical inspections of the place, and twice I came across tigers lying about in the vicinity of the enclosure – they probably came to challenge the lions!

We did not let out all the lions at the same time, but they were released in pairs. The first pair which was let out in August 1920 gave us no trouble, but vanished in the wilderness, But when the second pair was let out, the animals came back again and made their home outside the enclosure. They caused great alarm among the men who went there with a supply of their food. They attacked and snatched away the buffalo from their hands. Fortunately they did not kill any man but they simply took the buffalo and started feeding on it there and then.

On getting this news we got rather worried; so the next day we went there in a party and drove them away from the enclosure. Since there were some more lions left in the enclosure a regular supply had to be sent for their feed. The next day when the shikaris went with a fresh buffalo they found the male lion lying dead with his body badly mutilated, showing that he had been killed by a tiger. The lioness was not seen anywhere in the vicinity. What had apparently happened was that this pair on being driven away must have come across some tiger in the jungle who must have killed the lion, and the lioness must have escaped.

The third, fourth and fifth pairs gave us no trouble, but when the sixth and last pair was let out after two months they proved most troublesome. They adopted the easiest method for getting their food. The forest in this part is very thinly populated having no big villages but just a few scattered hamlets. The poor villagers do not possess any fire-arms. The pair of lions made the habit of going to these hamlets and helping

"Lion (mane not fully developed) from Eastern Asia, with Lioness".
*The English Cyclopaedia, Natural History*, Vol. II, 1854.

themselves to any cattle they could kill and eat on the spot. The villagers, to protect their animals, built stronger fences. The next time the pair visited the village, they could not get through those fences and therefore they killed a man instead and devoured him. As soon as this news was brought to us we rushed to the spot and destroyed the animals.

Most of the five pairs that vanished into the wilderness went a long way east and south. A few cases came to my knowledge of these lions having been actually shot near Panna and Jhansi in the east, and some at Kotah in the south. The late Maharaja of Baria shot one of them a few years ago along the bank of Kunoo River in Madhya Bharat.

I was glad to read in the newspapers that there is a proposal to re-introduce the Indian lion from the Gir forest into some other parts of our country, so that the species may not get extinct. If this idea is under serious contemplation, I suggest that the authorities should select isolated forests in which there are no tigers. Rajasthan is one of the suitable provinces where one can find such isolated jungles. It is most desirable to make this experiment, because very few Indian lions are left in the world, and if they die the species will vanish with them.

There is a great difference between the habits of these two animals as well. A lion uses his paws to strike his adversary, whereas the tiger uses them mainly for holding down his victim. Lions live in a "pride" consisting of a large family, whereas the habit of the tiger in this respect is just the opposite. Lions do their hunting by team work which tigers rarely do. The lion is comparatively weaker but bolder, and he is not half as cunning as the tiger. If a tiger is accompanied by a tigress and cubs it is the tiger who tackles the kill first, and he has his fill before allowing any member of this family to touch the food. But in the case of the lion, and also the panther, it is the female who does the killing and eating, while the male joins her later on. To put it in nut-shell a tiger has more of the Indian habit in this respect than the other animals!

In conclusion I must state that our implanting experiments were more of a success than a failure. The very fact that H.H. the Maharao of Kotah, and the Maharajas of Panna and Baria have shot these lions in comparatively recent years, suggests the possibility that they may still be surviving in remote areas away from the haunts of the tiger.

KESRI SINGH, COLONEL

Narain Niwas, Jaipur (Rajasthan),
26th October 1955.
*JBNHS*, Vol. LIII.

## Footprints of "Snowman"

I was on the fourth trip to Rupkund on September 16, 1956, and was camping in the rock-shelter of Baguva-vasa, 3½ miles before reaching Rupkund. It began to snow from 2.30 p.m. with terrific thunder-claps at intervals, and fine hail bigger than mustard seed fell. By 5 p.m. the snowfall stopped; the sky became completely clear by 6 p.m. and there was bright moonlight. The depth of snow was 4 to 6 in. in front of the rock-shelter, but it was less towards Balpa-Sulera, a hundred yards on the east of Baguva-vasa and on the windward side.

On September 17, 1956 at about 4 a.m. the whole region was enveloped in thick mist; at 4.30 a.m. it began to snow intermittently; at about 9 a.m. there was a hailstorm for about 15 minutes. Thereafter the sky began to clear up, and the sun could be seen shining on the neighbouring hilltops, but there was mist still here and there though Rupkund was seen clearly.

Leaving the luggage in the rock-shelter, I started with my two porters at 10 a.m. towards Rupkund. We had hardly proceeded a hundred yards to the place called Balpa-Sulera, with Nanda Ghunti, Trisul, and Chananiya Kot peaks, and Rupkund in front of us, when I suddenly saw some footprints on the ground. Casually I enquired of my porters if there were any panthers or hyenas in that region; they said that there were "lakad-baggha". On a close examination of the footprints, after removing my goggles, I found them to be like those of a human being. The elderly porter immediately burst out: "Footprints of Chananiyas, the doliyas or palanquin-bearers of Nanda Bhagavati".

The trail of footprints was seen coming from the direction of Rupkund and going towards the ruins of Balpa-Sulera, and then up over the Baguva-vasa rock-shelter. I could not follow the trail, since the animal had travelled over a steep track from one rock to another, and since the footprints were not in one plane. The footprints measured 5¾" x 2¾" and one to two inches in depth, on fresh snow. The impressions were quite clear and fresh with all the five toes and heel distinctly seen; black spots of bare ground could also be seen in some footprints. They were just like those of a human boy. The animal must have passed that way after the fall of the last hail before the mist cleared at 9 a.m., since there could not be seen any hailstones in the footprints, whereas small pearl-like hail was lying about on the surrounding snow. So, the animal must have passed that way at the most an hour before I saw its footprints. At places there were three, or rather 2½ footprints; and so the animal must have been on all fours, at least at those places. Unfortunately I could not pursue the trail for more than 50 yards on either side, first because the trail led up a steep ascent and secondly because the porters were in great haste and fear as it had begun to snow again at about 10.45 a.m. Reluctantly I had to return to Wan, giving up all chances of a fuller investigation.

I am of opinion that the footprints I saw could well be those of a baby Brown Bear, which might have come down from Rupkund side or Gingtoli plain. It had snowed heavily on the previous day as well as the following morning and the whole slope from Kailvavinaik to Rupkund was one continuous white sheet. The animal might have come down this slope in search of food. If not a Brown Bear it may have been an ordinary Black Bear from the near-by jungles, which are situated within a radius of four miles on all sides.

It may be mentioned in this connection that in the year 1905 Lord Curzon, Viceroy of India, visited Jetha Kharik, 1½ miles east of Ali Khal and about 6 miles from Baguva-vasa (where I found the footprints), for shooting brown bear. A road was specially constructed at that time from Wan to Ali Khal, which still exists and which has been repaired by Shramadan. So it would not be surprising for the footprints I noticed near Balpa-Sulera, at a height of about 14,000 ft. above sea level, to be those of a brown bear cub, with its mother sitting somewhere near by; or perhaps even of a black bear, which is very common in the neighbouring jungles. I came across a solitary male bharal near Chedi-nag midway between Baguva-vasa and Rupkund on August 25, 1956, on my first trip to Rupkund.

When I was on my fifth trip to Rupkund, on October 7, 1956, I came across a trail of footprints, a little beyond Patar-Nachauniya (about a mile before reaching Baguva-vasa), which were round at one end and tapering at the other. They were 4 in. long, 3 in. wide, and ⅞ in. deep. They were found on snow which had fallen four or five days previously. The upper part of the snow was encrusted so hard that my

**Small Minivet >**
*Pericrocotus cinnamomeus*
Adult male chiefly black, grey, and orange-crimson. Female and young male with no black on head and with yellow largely replacing the red. Arboreal. Bird of the plains, common in small parties fluttering in tree canopies. Affects gardens, groves, and light deciduous jungle in flocks. A feeble musical *swee-swee* call while hunting and while on the wing. Found throughout the plains and lower hills of the Indian Union, Bangladesh, Pakistan, Sri Lanka, Myanmar.

**Small Minivet (Little Minivet)** *Pericrocotus cinnamomeus* (Linnaeus)
*Birds of Asia*, Vol. II, Parts VII–XII, by John Gould, 1855–60. Painted by John Gould & Henry C. Richter.

Sponsored by Jewelex India Pvt. Ltd.

feet were not sinking in it at all but were skidding at several places. Several of the footprints were found in the middle of human footprints, that were left two days previously by an advance party who had gone up to Baguva-vasa. On enquiry from the elderly people of Wan village I was told that the footprints might be those of a tharuva (Snow Leopard) or a lakad-baggha (hyena).

Some villagers of Wan reported that the footprints of Chananiya (footprints like those of human beings) were noted occasionally in winter at Bagchho and Kukin Khal (3½ and 8 miles respectively from Baguva-vasa). So the strange footprints of the so-called snowman are apparently known in the Rupkund region from long years, though the villagers do not suspect them to belong to a bear. They believe them to be those of Chananiyas, whose abode is said to be the peak Chananiya Kot, situated on the northern side of Rupkund.

In the paragana of Danpur of Almora District and in Badhan and adjoining paraganas of Garhwal District, Chananiya is a Vana Devata or deity of the forests. She is said to have the feet reversed, i.e. toes pointing backward and heel in front. So, this deity is also called Ediya (heeled one). When lone women go to the jungle for cutting grass they are said to be frightened and affected by the evil influences of this deity. To get rid of these evil effects, the afflicted persons propitiate this deity. So far as my knowledge goes, there are two shrines dedicated to this deity, called Ediyaka Than – one between Gwaldam and Garur in Almora District, and the other near about Karnaprayag in Garhwal District.

Langurs or Blackfaced Monkeys are very timid and I never heard of one biting a man, excepting perhaps at Jagannath where they are fed freely by pilgrims and where they often become bold enough to snatch away food from their hands. No doubt Redfaced Monkeys attack man and even bite, assault, and injure very badly. I have never seen the Blackfaced Monkeys beyond the tree line or at heights above 10,000 feet; as such there seems absolutely no possibility whatsoever of either the Redfaced Monkey or the Blackfaced Langur having left the footprints at Baguva-vasa, which is at an altitude of 14,000 ft. Besides, the footprints of a langur monkey are in fours, quite different and distinct from human footprints. The footprints I saw were just like those of a human being. Monkeys and langurs do not go beyond Wan and Sutol in the Rupkund region.

Two Norwegian engineers Age Thorberg and Jan Forstis also had an encounter with two langur-like animals, one of which is alleged to have bitten Forstis.

SWAMI PRANAVANANDA, F.R.G.S.

Almora, U.P., 6th November 1956.
*JBNHS*, Vol. LIV.

**Scarlet Minivet >**
*Pericrocotus flammeus*
An exclusively arboreal bird, male black and scarlet; female and young male grey, olive-yellow, and yellow. Affects well wooded country and evergreen forests. In winter seen in small flocks which travel through treetops searching for insects. Like other minivets, these birds flit from tree to tree in follow-my-leader fashion, their bright colours glinting in the sunlight and their cheery calls enhancing the pleasure of meeting the flock. Resident, practically throughout India but mainly in the Himalayas from Jammu & Kashmir east to Arunachal Pradesh; locally in the hills of India up to about 2,000 m, Bangladesh and Sri Lanka.

Scarlet Minivet *Pericrocotus flammeus* (Forster)
*Birds of Asia*, Vol. II, Parts VII–XII, by John Gould, 1855–60. Painted by John Gould & Henry C. Richter.

Courtesy Lataben & Sevantibhai Parikh, Jagruti & Premal Parikh, Mumbai

# Bibliography

Aitken, E.H.
*Behind the Bungalow.* 1889.
*The Common Birds of Bombay.* 1894.
*The Naturalist on the Prowl.* 1894.
*The Tribes on my Frontier.* 1898.
*The Gazetteer of the Province of Sind.* 1907.
*Concerning Animals and Other Matters.* 1914.

Ali, Sálim.
*The Birds of Kutch.* 1945.
*Indian Hill Birds.* 1949.
*The Birds of Travancore and Cochin.* 1953.
  Third revised edition published as *Birds of Kerala.* Revised by R. Sugathan, edited by J.C. Daniel. 1999.
*The Birds of Sikkim.* 1962.
*Field Guide to the Birds of the Eastern Himalayas.* 1977.
*The Book of Indian Birds.* Thirteenth edition, 2002.

Ali, Sálim, and S. Dillon Ripley.
*Handbook of Birds of India and Pakistan.* 10 Vols. 1968–74.
*A Pictorial Guide to the Birds of the Indian Subcontinent.* 1983.

Allen, Charles.
*Kipling's Kingdom.* 1987.

Almeida, M.R.
*Flora of Maharashtra.* 4 Vols. 1996–2004.

Baker, E.C. Stuart.
*The Birds of North Cachar.* 1901–02.
*The Indian Ducks and their Allies.* 1908.
*Indian Pigeons and Doves.* 1913.
*The Game Birds of India, Burma and Ceylon.* 3 Vols. 1921–30.
*The Fauna of British India. Birds.* 8 Vols. Second edition, 1922–30.
*The Nidification of Birds of the Indian Empire.* 4 Vols. 1932–35.

Baker, Lt.-Col. H.R., and C.M. Englis.
*The Birds of Southern India, including Madras, Malabar, Travancore, Cochin, Coorg, and Mysore.* 1930.

Baldwin, Captain J.H.
*The Large and Small Game of Bengal and the North-western Provinces of India.* 1876.

Balfour, Surgeon General Edward.
*The Cyclopaedia of India and of Eastern and Southern Asia.* 3 Vols. 1885.

Barnes, Lieut. Edwin H.
*Handbook to the Birds of Bombay Presidency.* 1885.

Bates, Capt. R.S.P.
*Bird-life in India.* 1931.

Bates, R.S.P., and E.H.N. Lowther.
*Breeding Birds of Kashmir.* 1952.

Benthall, A.P.
*The Trees of Calcutta and its Neighbourhood.* 1946

Blanford, W.T.
*The Fauna of British India Including Ceylon and Burma (Mammalia).* 1888–91.
*The Fauna of British India (Birds).* Vol. III. 1895. Vol. IV. 1898.

Blatter, Ethelbert, and W.S. Millard.
*Some Beautiful Indian Trees.* 1937.

Brander, Dunbar, A.A.
*Wild Animals in Central India.* Third impression, 1931.

Brett, E.A. de.
*Central Provinces Gazetteers, Chhattisgarh Feudatory States.* 1909.

Brown, C.
*Central Provinces & Berar District Gazetteers, Akola District.* 1910.

Burton, Brigadier-General R.G.
*Sport & Wildlife in the Deccan.* 1926.

Butler, John.
*Travels in Assam.* 1855.

Campbell, James.
*Gazetteer of the Bombay Presidency.*
Vol. II *Surat & Broach.* 1877.
Vol. III *Kaira & Panchmahals.* 1879.
Vol. IV *Ahmedabad.* 1879.
Vol. V *Kutch, Palanpur & Mahikantha.* 1880.
Vol. VI *Rewa Kantha, Narukot, Cambay, & Surat States.* 1880.
Vol. VIII *Thana.* Part II 1882.
Vol. X *Ratnagiri & Savantvadi.* 1880.
Vol. XI *Kolaba & Janjira.* 1883.
Vol. XII *Khandesh.* 1880.
Vol. XV *Kanara.* Part I, 1883. Part II, 1883.
Vol. XVI *Nasik.* 1883.
Vol. XVII *Ahmadnagar.* 1884.
Vol. XVIII *Poona.* Parts I to III, 1885.
Vol. XIX *Satara.* 1885.
Vol. XX *Sholapur.* 1884.
Vol. XXI *Belgaon.* 1884.
Vol. XXII *Dharwar.* 1884.
Vol. XXIII *Bijapor.* 1884.
Vol. XXIV *Kolhapur.* 1886.

Campbell, Captn. Walter.
*The Old Forest Ranger or Wild Sports of India on the Neilgherry Hills in the jungles and on the planes.* 1842.

Casserly, Lieut.-Colonel Gordon.
*Dwellers in the Jungle.* 1925.

Colthrust, Ida.
*Familiar Flowering Trees of India.* 1924.

Cowen, D.V.
*Flowering Trees and Shrubs in India.* Third edition, 1957.

Crooke, W.
*Folklore of Northern India,* in two volumes. 1896.

Daniel, J.C.
*The Book of Indian Reptiles and Amphibians.* 2002.

Delacour, Jean.
*The Pheasants of the World.* 1957.

Deoras, P.J.
*Snakes of India.* 1965.

Desmond, Ray.
*The European Discovery of Indian Flora.* 1992.

Dharmakumarsinhji, R.S.
*Birds of Saurashtra.* 1954.

Douglas, James.
*Bombay and Western India.* Vols. I & II. 1893.

Drake-Brockman, D.L.
*District Gazetteers of the United Provinces of Agra and Oudh. Banda.* 1909. *Jalaun.* 1909. *Jhansi.* 1909.

Dunn, Sara H.
*Sunny Memories of an Indian Winter.* 1898.

Eardley-Wilmot, S.
*The Life of a Tiger.* 1911.

Elliot, Commander Robert.
*Views in India, China, and the Shores of the Red Sea.* With Descriptions by Emma Roberts. Vol. I. 1835.

Enthoven, R.E.
*The Folklore of Bombay.* 1924.

Finn, Frank.
*Indian Sporting Birds.* 1915
*Sterndale's Mammalia of India.* 1929.

Forbes, James.
*Oriental Memoirs.* 4 Vols. 1812–13.

Forsyth, Capt. J.
*The Highlands of Central India.* 1889.

Gee, E.P.
*The Wildlife of India.* 1964.

Gould, John.
*A Century of Birds from the Himalaya Mountains.* 1832.
*The Birds of Asia.*
Vol. I, Parts I to VI, 1850–54.
Vol. II, Parts VII to XII, 1855–60.
Vol. III, Parts XIII to XVIII, 1861–66.
Vol. IV, Parts XIX to XXIV, 1867–72.
Vol. V, Parts XXV to XXX, 1873–77.
Vol. VI, Parts XXXI to XXXV, 1869–73.

Gouldsbury, C.E.
*Tiger Slayer by Order.* 1916.

Greenway, James C.
*Extinct and Vanishing Birds of the World.* 1958.

"**Peasants at a Well in Hindostan**". Sketched by Baron de Montalembert, 1807.
*Oriental Memoirs*, Vol. III, by James Forbes, 1813.
Note the leather water-bag loaded on the bullock; this was traditionally used by Indian *bhistis* or water-carriers in the olden days.

Grimmett, Richard, Carol and Tim Inskipp.
*Birds of the Indian Subcontinent.* 1998.

Grindlay, Capt. Robert Melville.
*Scenery, Costumes and Architecture Chiefly on the Western Side of India.* 1826.

Gruning, John F.
*Eastern Bengal and Assam District Gazetteers – Jalpaiguri.* 1911.

Guy, John, and Deborah Swallow.
*Arts of India, 1550–1900* (A Victoria and Albert Museum, London publication). 1999.

H.A.L.
"The Old Shekarry", *Sport in Many Lands.* 1890.

Hamilton, General Douglas.
*Records of Sport in South India, Chiefly on the Annamullay, Nielgherry and Pulney Mountains.* 1892.

Heber, Rev. Reginald (Lord Bishop of Calcutta).
*Narrative of a Journey through the Upper Provinces of India, Calcutta to Bombay, 1824–25, an account of a journey to Madras and Southern Provinces, 1826.* In two volumes, 1843–44.

Horsfield, Thomas, and Frederic Moore.
*A Catalogue of the Birds in the Museum of the Hon. East India Company.* In two volumes. 1854–58.

Hume, Allen O.
*Stray Feathers.* Vols. I to XI. 1873–99.
*The Nests and Eggs of Indian Birds.* Vols. I to III. 1889–90.

Hume, Allen, and C.H.T. Marshall.
*The Game Birds of India, Burma & Ceylon.* Vols. I to III. 1879–81.

Hutson, Major-General H.P.W.
*The Birds about Delhi.* 1954.

Jerdon, T.C.
*The Mammals of India.* 1862.
*The Birds of India.* Vol. I, 1862. Vol. II, 1863. Vol. III, 1864.

Khory, Rustomjee Naserwanjee.
*The Bombay Materia Medica.* 1887.

Kipling, Rudyard.
*The Jungle Book.* 1895.
*The Second Jungle Book.* 1895.

Kirkman, F.B., and F.C.R. Jourdain.
*British Birds.* 1930.

Knight, Charles.
*The English Cyclopaedia, Natural History.* Vols. I to III. 1854–55.

Kothari, Ashok S., and B.F. Chhapgar, eds.
*Sálim Ali's India.* 1996.

Low, C.L.
*Central Provinces District Gazetteers, Balaghat District.* 1907.

Lydekker, Richard, Sir Harry Johnston, and Prof. J.R. Ainsworth-Davis.
*Harmsworth Natural History.* Vol. I. 1910.

Macintyre, Major-General Donald.
*Hindu-Koh.* 1891.

Mackintosh, L.J.
*Birds of Darjeeling and India.* In 2 parts. 1915.

Malcolm. Maj.-Gen. Sir John.
*A Memoir of Central India including Malwa.* Vols. I & II. 1824.

Murray, James.
*Plants and Drugs of Sind.* 1881.

Nairne, Alexander K.Y.D.
*The Flowering Plants of Western India.* 1894.

Nelson, A.E.
*Central Provinces District Gazetteers. Raipur District.* 1909. *Durg District.* 1910.

Nevill, H.R.
*District Gazetteers of the United Provinces of Agra and Oudh. Agra.* 1909. *Jaunpur.* 1908.

Oates, Eugene W.
*The Fauna of British India (Birds).* Vols. I & II. 1889–90.

O'Malley, L.S.S.
*Bengal District Gazetteers. 24-Parganas.* 1914. *Murshidabad.* 1914.

Prater, S.H.
*The Book of Indian Animals.* Fourth impression, 1993.

Raghu Vira, Dr. and K.N. Dave.
*Indian Scientific Nomenclature of Birds of India, Burma and Ceylon.* 1949.

Randhawa, M.S.
*Paintings of the Babur Nama.* (A National Museum, Delhi publication). 1983.

Ripley II, Sidney Dillon.
*A Synopsis of the Birds of India and Pakistan.* Second edition, 1982.

Royle, J.Forbes.
*Illustrations of the Botany and other branches of Natural History of the Himalayan Mountains and the Flora of Cashmere.* Vol. I. 1839.

Russel, R.V., ed.
*Central Provinces District Gazetteers, Saugor District.* 1906. *Betul District.* 1907. *Seoni District.* 1907. *Bhandara District.* 1908.

Sahani, K.C.
*The Book of Indian Trees.* 2000.

Seton, Grace Thompson.
*"Yes Lady Saheb" A Woman's Adventurings in Mysterious India.* 1925.

"Silver Hackle".
*Indian Jungle-lore and the Rifle.* 1929.

Singh, Colonel Kesari.
*The Tiger of Rajasthan.* 1959.

Sterndale, Robert Armitage.
*Natural History of the Mammalia of British India and Ceylon.* 1884.
*Denizens of the Jungles.* 1886.

Stewart, Col. A.E.
*Tiger and Other Game.* 1927.

Stockley, Lieut.-Col. C.H.
*Big Game Shooting in the Indian Empire.* 1928.

Temple, Sir Richard.
*Men and Events of my Time in India.* 1882.

Tennent, Sir J. Emerson.
*Sketches of the Natural History of Ceylon.* 1861.

Troup, R.S.
*The Silviculture of Indian Trees.* Vols. I to III. 1921.

Vas, J.A.
*East Bengal and Assam District Gazetteers. Rangpur.* 1911.

Wall, Major F.
*The Poisonous Terrestrial Snakes of our British Indian Dominions.* Third edition, 1913.

Walton, H.G.
*District Gazetteer of The United Provinces of Agra & Oudh. Dehra Dun.* 1911.

Webster, J.E.
*Eastern Bengal District Gazetteers. Tippera.* 1910.

Whistler, Hugh.
*Popular Handbook of Indian Birds.* Fourth edition, 1949.

Williams, Major-General H.
*The Birds around Delhi.* 1954.

Williamson, Thomas.
*Oriental Field Sports.* 2 Vols. 1808.

Wood, The Rev. J.G.
*Routledge's Picture Natural History.* 1885.

Wynter-Blyth, M.A.
*Butterflies of the Indian Region.* 1957.

Yule, Col. Henry, and A.C. Burnell.
*Hobson-Jobson.* 1886.

Various issues of the *Journal of the Bombay Natural History Society*, *Bengal Sporting Magazine*, *Journal of the Bengal Natural History Society*, and *Journal of the Darjeeling Natural History Society*.

# Sponsors

ADITYA REALTORS LIMITED
Aditya Kanoria, Director
4, India Exchange Place, 3rd Floor,
Kolkata 700001, West Bengal, India
(Page 99)

ADVANCE COMPUTER SERVICES
PRIVATE LIMITED
Surendra J. Patel,
Chairman & Managing Director
Unit Nos. 201–204, Sector II,
Building 5, Millenium Business Park,
Mhape, Navi Mumbai 400709,
Maharashtra, India
(Page 140)

ASHOK KUMAR MEHRA & CO.
Chartered Accountants,
G1/G2, "Saujanya", Plot No. 527B,
16th Road, Khar (W), Mumbai 400052
Maharashtra, India
(Page 157)

ASIAN STAR CO. LTD.
114, Mittal Court – C,
Nariman Point, Mumbai 400021,
Maharashtra, India
(Page 54)

ASTROX CORPORATION
Dr Ajay P. Kothari, President & CEO
3500 Marlbroughway, College Park,
Maryland 20740, USA
(Pages 80, 156)

AVOCENT INDIA
Sudhir Seth, Managing Director
303, Samarpan Complex, New Link Road,
Chakala, Andheri (E), Mumbai 400099,
Maharashtra, India
(Page 33)

B.G. DESHMUKH
41, Buena Vista, Gen. J. Bhonsale Marg,
Mumbai 400021, Maharashtra, India
(Page 201)

BAJAJ AUTO LTD.
Mumbai-Pune Road, Akurdi,
Pune 411035, Maharashtra, India
(Page 51)

BHAILAL AMIN FOUNDATION
Amaltas Farm & Nursery,
Bhayli-Raipura Village Road, Bhayli,
District Vadodara 391410, Gujarat, India
(Page 183)

BHANSALI & CO.
Jitu Bhansali, Partner
640-646, Panchratna,
Mama Parmanand Marg, Opera House,
Mumbai 400004, Maharashtra, India
(Page 153)

BHARAT FLOORINGS & TILES
(MUMBAI) PVT. LTD.
Dilnavaz S. Variava, Chairperson
32, Mumbai Samachar Marg,
Next to Stock Exchange,
Fort, Mumbai 400023,
Maharashtra, India
(Page 143)

BLUE CROSS LABORATORIES
LIMITED
N.H. Israni, CMD
Peninsula Chambers, P O. Box 16360,
Lower Parel, Mumbai 400013,
Maharashtra, India
(Page 127)

CENTRAL BANK OF INDIA
Chander Mukhi, Nariman Point,
Mumbai 400021, Maharashtra, India
(Page 21)

CHANDRAKANT A. PATEL
Holiday Inn Express Hotel & Suites,
12439, Northwest Freeway,
Houston, Texas 77092, USA
(Page 111)

D. NAVINCHANDRA & COMPANY
212/214, Prasad Chambers, Opera House,
Mumbai 400004, Maharashtra, India
(Page 180)

DALLAS DIAMONDS CORPORATION
Nikin Mehta, President
5580 LBJ Freeway, #645,
Dallas, Texas 75240, USA
(Page 39)

DIAMONDSTAR
Chhotalal Premchand Shah, Partner
804, Panchratna, Opera House,
Mumbai 400004, Maharashtra, India
(Page 91)

DIANA RATNAGAR
4, Prince of Wales Drive, Wanowrie,
Pune 411040, Maharashtra, India
(Page 121)

DILIPKUMAR V. LAKHI
102/115, Prasad Chambers,
Opera House, Mumbai 400004,
Maharashtra, India
(Page 179)

DIMEXON DIAMONDS LIMITED
704, Raheja Chambers,
213, Nariman Point, Mumbai 400021,
Maharashtra, India
(Pages 93, 177)

DIWALIBEN MOHANLAL MEHTA
CHARITABLE TRUST
Khatau Mansion, Oomer Park,
95 K, Bhulabhai Desai Road,
Mumbai 400036, Maharashtra, India
(Page 163)

DR. REDDY'S LABORATORIES
LIMITED
G.V. Prasad, Vice Chairman & CEO
7-1-27 Ameerpet, Hyderabad 500016,
Andhra Pradesh, India
(Page 40)

GARDEN SILK MILLS LIMITED
Manek Mahal, 90, Veer Nariman Road,
Mumbai 400020, Maharashtra, India
(Page 113)

DR. GAYATRI UGRA
Bombay Natural History Society
(Page 77)

GODREJ INDUSTRIES LIMITED
Pirojshanagar, Eastern Express Highway,
Vikhroli, Mumbai 400079,
Maharashtra, India
(Pages 89, 185)

GOKUL ICE-CREAMS
Ganesh Kamath
Ajay Apartments, Saraswati Road, Santacruz (W),
Mumbai 400054, Maharashtra, India
(Page 132)

GOODLASS NEROLAC PAINTS LTD.
"Nerolac House", Ganpatrao Kadam Marg,
Lower Parel, Mumbai 400013,
Maharashtra, India
(Page 53)

GRINDWELL NORTON LTD.
Army & Navy Building, 148, M.G. Road,
Fort, Mumbai 400001, Maharashtra, India
(Page 109)

GUJARAT AMBUJA CEMENTS LTD.
122, Maker Chambers III, Nariman Point,
Mumbai 400021, Maharashtra, India
(Page 24)

HDFC
Housing Development Finance
Corporation Limited
Ramon House, H.T. Parekh Marg,
169, Backbay Reclamation, Churchgate,
Mumbai 400020, Maharashtra, India
(Page 48)

HIMMATSINGKA SEIDE LIMITED
2/1 Midford Gardens, M.G. Road,
Bangalore 560001, Karnataka, India
(Page 151)

HINDUSTAN EXPORT & IMPORT
CORPORATION PVT. LTD.
Kanwal K. Grover,
Chairman & Managing Director
Anand Bhavan, 348, Dr D.N. Road,
P.O. Box No. 1091,
Mumbai 400001, Maharashtra, India
(Page 63)

HSBC
The Hongkong & Shanghai Banking
Corporation Limited
52/60 Mahatma Gandhi Road,
Mumbai 400001, Maharashtra, India
(Page 36)

INDIAMCO
Satish D. Choksi, Chairman
602, Panchratna, Opera House,
Mumbai 400004, Maharashtra, India
(Page 62)

INDUSTRIAL MANUFACTURERS
525, Sayani Road, Prabhadevi,
Mumbai 400025, Maharashtra, India
(Page 159)

ION EXCHANGE (INDIA) LTD.
Tiecicon House, Dr. E. Moses Road,
Mahalaxmi, Mumbai 400011,
Maharashtra, India
(Page 16)

ITC LIMITED
Virginia House, 37, J.L. Nehru Road,
Kolkata 700071, West Bengal, India
(Page 193)

JEWELEX INDIA PVT. LTD.
Piyush Kothari, Chairman
605 Aman Chambers, 113 Queen's Road,
Mumbai 400004, Maharashtra, India
(Page 207)

JINDAL VIJAYANAGAR STEEL LIMITED
Jindal Mansion, 54, Dr G. Deshmukh
Marg, Mumbai 400026, Maharashtra, India
(Page 148)

KALPATARU TRUST
Hornby View Bldg., 1st Floor,
12, Rustom Sidhwa Marg,
Mumbai 400001, Maharashtra, India
(Page 147)

LONDON STAR DIAMOND COMPANY
(INDIA) PVT. LTD.
Kamlesh Jhaveri, Managing Director
1610/11 Prasad Chambers, Opera House,
Mumbai 400004, Maharashtra, India
(Pages 23, 191)

M. SURESH & CO.
Suresh Pethani, Partner
1212, Prasad Chambers, Opera House,
Mumbai 400004, Maharashtra, India
(Page 194)

MAHENDRA BROTHERS
611, Panchratna, Mama Parmanand Marg,
Mumbai 400004, Maharashtra, India
(Page 173)

MAHENDRA PREMCHANDBHAI SHAH
"Saprem", JVPD Scheme, Road No. 3,
Vile Parle (W), Mumbai 400056,
Maharashtra, India
(Page 38)

MAHINDRA & MAHINDRA LTD.
Anand Mahindra,
Vice Chairman & Managing Director
Gateway Building, Apollo Bunder,
Mumbai 400001, Maharashtra, India
(Page 59)

MAIZE PRODUCTS
Priyam B. Mehta, Managing Director
P.O. Kathwada, Ahmedabad 382430,
Gujarat, India
(Page 164)

MANJULA DHIRAJLAL SHROFF
Apna Cottage, Juhu-Tara Road,
Mumbai 400049, Maharashtra, India
(Page 64)

MANU M. PARPIA
72, Tenerife, Little Gibbs Road 2,
Malabar Hill, Mumbai 400006,
Maharashtra, India
(Page 150)

MASTEK FOUNDATION
106, SDF-IV, SEEPZ, Andheri (E),
Mumbai 400096, Maharashtra, India
(Page 43)

MSPL LIMITED
Shrenik N. Baldota, Executive Director
Baldota Bhavan, Maharshi Karve Road,
Mumbai 400020, Maharashtra, India
(Back Cover and Front Endpaper)

NAVDEEP CHEMICALS PVT. LTD.
166, Bora Bazar Street,
Fort, Mumbai 400001,
Maharashtra, India
(Page 176)

OOPAL DIAMOND
Ashish Doshi
6/15-B Sopariwala Estate,
Prasad Chambers Compound,
Opera House, Mumbai 400004,
Maharashtra, India
(Page 119)

P. SEVANTILAL & CO.
Premal Parikh, Partner
403, Panchratna,
Opera House, Mumbai 400004,
Maharashtra, India
(Page 209)

PAKONA ENGINEERS (I) PVT. LTD.
Ashok J. Kothari, Chairman &
Managing Director
22-D, Wadia Charities Bldg., 1st Floor,
S.A. Brelvi Road, Fort,
Mumbai 400023, Maharashtra, India
(Page 78)

PARAM PREET SINGH
B-8, Namdhari Chambers,
9/54, D.B. Gupta Road, Karol Bagh,
New Delhi 110005, India
(Page 32)

PARIKH FOUNDATION
Kavinbhai Parikh, Chairman
611, Panchratna, Mama Parmanand Marg,
Mumbai 400004, Maharashtra, India
(Page 115)

PHEROZA & JAMSHYD GODREJ
"The Trees", 1st Floor,
40-D, Ridge Road, Malabar Hill,
Mumbai 400006, Maharashtra, India
(Page 75)

PIROJSHA GODREJ FOUNDATION
Godrej Bhavan, Home Street,
Fort, Mumbai 400001, Maharashtra, India
(Front Cover)

PITTI LAMINATIONS LIMITED
6-3-648/401, Padmaja Landmark, 4th Floor,
Somajiguda, Hyderabad 500082,
Andhra Pradesh, India
(Page 84)

POULTRY DEVELOPMENT
PROMOTION COUNCIL
"Venkateshwara House",
3-5-808/1, Hyderguda,
Hyderabad 500029,
Andhra Pradesh, India
(Pages 83, 202)

PUSHPA KIRTILAL BHANSALI
3, Soona Mahal, Marine Drive,
Mumbai 400020, Maharashtra, India
(Page 67)

RAHULKUMAR BAJAJ CHARITABLE
TRUST
Bajaj Bhavan, 2nd Floor,
226, Nariman Point, Mumbai 400021,
Maharashtra, India
(Page 66)

RAIKA & NAVROZE GODREJ
"The Trees", 1st Floor,
40-D, Ridge Road, Malabar Hill,
Mumbai 400006, Maharashtra, India
(Page 104)

RATI SHAH
Ahimsa of Texas,
1720E, Jeter Road, Bartonville,
Texas 76226, USA
(Page 189)

RENAISSANCE DIAMOND
CORPORATION
Raj Shekhar Parikh, President
424, Madison Avenue, New York,
NY 10017, USA
(Page 169)

RISHAD NAOROJI
Godrej Bhavan, Home Street,
Fort, Mumbai 400001, Maharashtra, India
(Pages 44, 81)

RUBY MADAN
15, Pankaj Mahal, Churchgate Reclamation,
Mumbai 400020, Maharashtra, India
(Page 128)

SATISH L. CHULLANI
"Mohini", 2nd Floor, 18th Road,
Khar (W), Mumbai 400052,
Maharashtra, India
(Page 29)

SHAKIR RANDERIAN
"Gulabi Block", Commerce House,
2, Ganesh Chandra Avenue,
Kolkata 700013, West Bengal, India
(Page 58)

SREEKANT S. MEHTA
Fairlands, 256, Advaita Ashram Road,
Salem 636016, Tamil Nadu, India
(Page 20)

STATE BANK OF INDIA
State Bank Bhavan, C-6, G Block,
Bandra-Kurla Complex, Bandra (E),
Mumbai 400051, Maharashtra, India
(Back Endpaper)

SUNIL DHAROD
5108, Silver Lake Drive,
Plano, Texas 75093, USA
(Page 28)

SUNSOKO
Bajaj Bhavan, 2nd Floor,
Jamnalal Bajaj Marg,
226, Nariman Point, Mumbai 400021,
Maharashtra, India
(Page 103)

SURESH S. KOTHARI
155, Damarest Avenue,
Englewood Cliffs,
New Jersey 07632, USA
(Page 57)

TATA CHEMICALS LIMITED
Bombay House,
24 Homi Mody Street, Fort,
Mumbai 400001, Maharashtra, India
(Page 101)

TATA CONSULTANCY SERVICES
LIMITED
Bombay House,
24, Homi Mody Street, Fort,
Mumbai 400001, Maharashtra, India
(Page 131)

TATA SONS LIMITED
Bombay House,
24 Homi Mody Street, Fort,
Mumbai 400001, Maharashtra, India
(Pages 71, 136)

THE INDIAN HOTELS COMPANY
LIMITED
The Taj Mahal Palace & Tower,
Apollo Bunder, Mumbai 400001,
Maharashtra, India
(Page 74)

THE RAJA BAHADUR MOTILAL
POONA MILLS LTD.
Shridhar Pittie, Executive Director
Hamam House, Ambalal Doshi Marg,
Mumbai 400023, Maharashtra, India
(Page 110)

THE TATA POWER COMPANY
LIMITED
Bombay House, 24 Homi Mody Street,
Fort, Mumbai 400001, Maharashtra, India
(Page 45)

TOLANI SHIPPING CO. LTD.
10-A, Bakhtawar, Nariman Point,
Mumbai 400021, Maharashtra, India
(Page 171)

TOYOTA KIRLOSKAR MOTOR PVT. LTD.
Plot No. 1, Bidadi Industrial Area,
Ramnagar Taluk,
Bangalore (Rural) Dist. 562109,
Karnataka, India
(Page 47)

UNITED SHIPPERS LIMITED
Sevantilal J. Parekh, Chairman
United India Building, Sir P.M. Road,
Mumbai 400001, Maharashtra, India
(Page 34)

UTI MUTUAL FUND
UTI Tower, G Block, Bandra-Kurla
Complex, Bandra (E),
Mumbai 400051, Maharashtra, India
(Page 95)

VASANT J. SHETH MEMORIAL FOUNDATION
Asha Sheth, Chairperson
The Great Eastern Shipping Co. Ltd.,
Energy House, 81 D N Road,
Mumbai 400001, Maharashtra, India
(Pages 97, 135)

VENUS JEWEL
Ramniklal Premchand Shah, Partner
902, Panchratna, Opera House,
Mumbai 400004, Maharashtra, India
(Page 117)

VIDHI, SHAIVI, MAAHIR, VEER & SHAAN
C/o Hansa Kothari, 10, Sushma,
Linking Road Extension, Santacruz (W),
Mumbai 400054, Maharashtra, India
(Page 145)

YASHWANT DATTATREYA JOSHI
2/1 Chandra Shekhar Society,
S.N. Marg, Andheri (E),
Mumbai 400069, Maharashtra, India
(Page 124)

## DETAILED CAPTIONS FOR JACKET, ENDPAPERS, AND TITLE PAGE

### FRONT COVER
Black-capped Kingfisher
*Halcyon pileata* (Boddart)
*Birds of Asia*, Vol. III, Parts XIII–XVIII, by John Gould, 1861–66. Painted by John Gould & Henry C. Richter.

A large kingfisher with a big coral bill, a velvety black cap separated from dark mantle by a prominent white collar on hind neck. Upper plumage looks black until sunshine transforms it to brilliant purple-blue. Readily identified also by its pale rusty under-parts including under-side of wings. During flight a large white patch in wing is clearly visible. Sexes alike. Seen singly near the seacoast and tidal rivers, frequently ascending along these for a considerable distance inland in forested country. Mangrove swamps bordering tidal creeks are favourite haunts. Call: a cackling laugh like the White-breasted Kingfisher's, but shriller and quite distinctive. Nest: a tunnel excavated in the earth-bank of a river or creek ending in a widened egg-chamber. Distribution: practically the entire coastline of India south of Mumbai, Andaman and Nicobar Islands, Bangladesh, Sri Lanka, Myanmar.

Remembering the late SOHRABJI PIROJSHA GODREJ
(former Vice -President, BNHS)
whose insatiable enthusiasm for nature will be long-remembered and admired.
Courtesy Pirojsha Godrej Foundation

### BACK COVER
Beautiful Nuthatch *Sitta formosa* Blyth
*Birds of Asia*, Vol. I, Parts I–VI, by John Gould, 1850–54. Painted by John Gould & Henry C. Richter.

A large, handsome, black and blue nuthatch. Crown, nape, and mantle black streaked with cobalt-blue. Sides of neck and back black streaked with bluish white. Rest of upper-parts blue. Flight and tail feathers black edged with blue, tail tipped bluish white. Forehead, sides of head, chin, and throat white shading into the dull chestnut lower plumage. Sexes alike. Keeps in parties of 4 or 5. Its actions and behaviour while creeping up and down boughs and tree trunks in search of food resembles those of a woodpecker. Call: typical of nuthatches, lower and sweeter in tone. Resident, very rare and local, recorded in E. Himalayas and NE hill states between 300 and 2,100 m.

Courtesy Shrenik N. Baldota, MSPL Limited

### FRONT ENDPAPER
"Approach to the Bore Ghaut". Drawn on the spot by William Westall, A.R.A. from a painting by Lt.-Col. Johnson, coloured by J.B. Hogarth, engraved by T. Fielding. From *Scenery, Costumes and Architecture Chiefly on the Western Side of India*, by Captn. Robert Melville Grindlay, 1826.

Courtesy Shrenik N. Baldota, MSPL Limited

### BACK ENDPAPER
"Fortress of Bowrie in Rajputana". Drawn by William Westall, A.R.A. from a drawing by Capt. Auber, engraved by C. Bentley. From *Scenery, Costumes and Architecture Chiefly on the Western Side of India*, by Captn. Robert Melville Grindlay, 1826.

Courtesy State Bank of India
Always Caring for Nature

### TITLE PAGE
Sclater's Monal
(Mishmi Monal Pheasant)
*Lophophorus sclateri* Jerdon
*Birds of Asia*, Vol. V, Parts XXV–XXX, by John Gould, 1873–77. Painted by John Gould & William Hart.

Male similar to Himalayan Monal, with metallic green, blue, and bronze coloration to head and upper-parts, and velvety-black under-parts; distinguished by crown having tuft of metallic bluish-green short curly feathers instead of upstanding crest. Tail, cinnamon with a broad white terminal band. All the metallic colours are less bright than the Himalayan Monal. Female, brown mottled and streaked with paler and darker brown, and with conspicuous greyish-white rump. Tail blackish, tipped with white and with narrow whitish bars. Resident, rarely found, in subalpine zone of Arunachal Pradesh (Kameng, Subansiri, Siang, and Luhit frontier divisions), between 3,000 and 5,000 m; in conifer forest with dense rhododendron undergrowth. Call: a ringing whistle. A Sclater's Monal was found in Mishmi Hills of Arunachal in 1870 and was sent to London Zoo the same year. It lived there for a year and eight months. Jerdon tells us that the bird was very tame and ate readily from the hand. He fed it on rice and corn, and it was especially fond of lettuce and cabbage leaves (Jean Delacour, *The Pheasants of the World*).

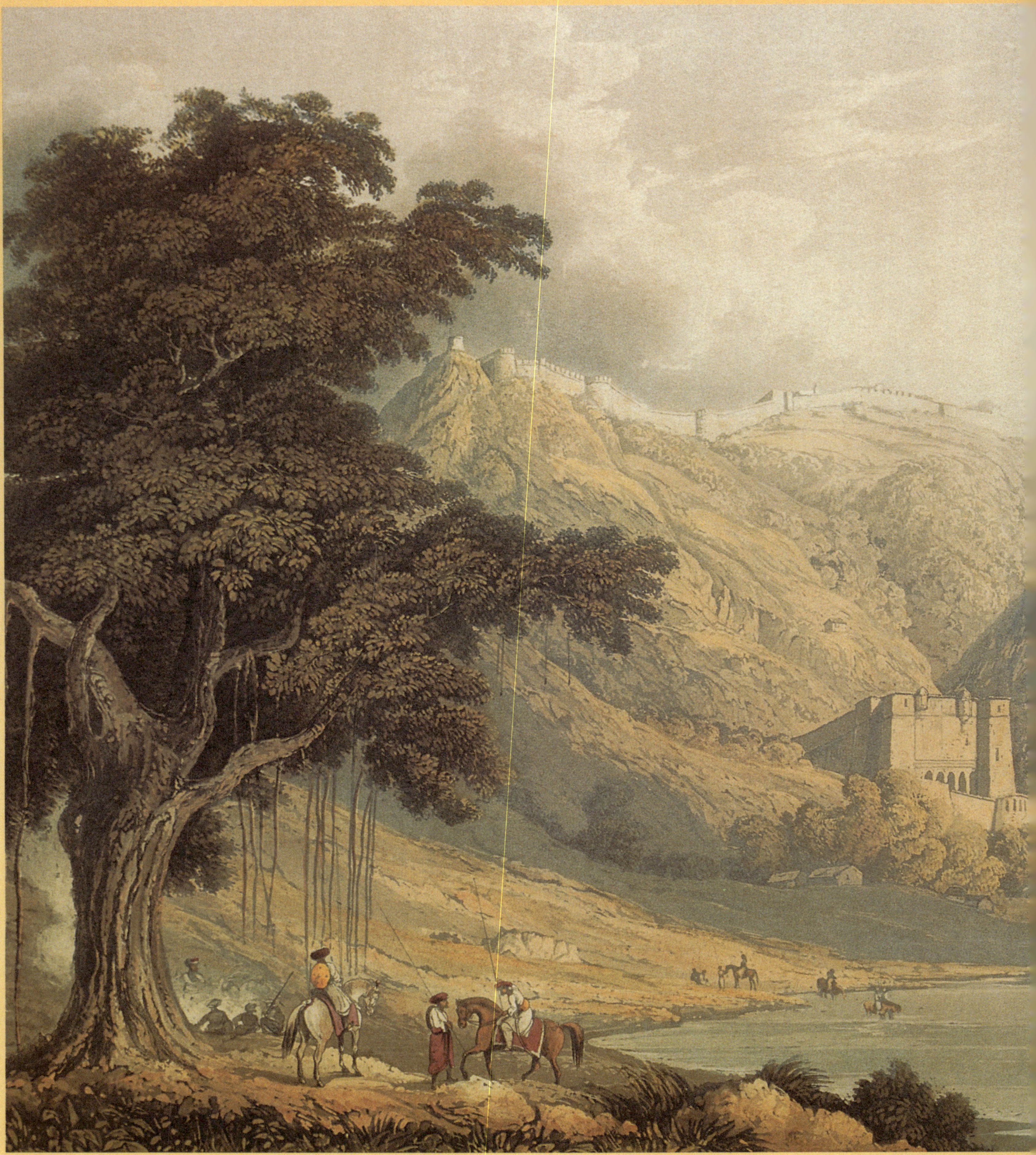